摩天大楼设计

[美] 盖特可事务所 编著
齐梦涵 译

广西师范大学出版社
· 桂林 ·
images Publishing

图书在版编目(CIP)数据

摩天大楼设计/美国盖特可事务所编著;齐梦涵译. —桂林:广西师范大学出版社,2018.5
ISBN 978-7-5598-0722-9

Ⅰ.①摩… Ⅱ.①美… ②齐… Ⅲ.①超高层建筑-建筑设计-作品集-美国-现代 Ⅳ.①TU972

中国版本图书馆CIP数据核字(2018)第052110号

出 品 人:刘广汉
责任编辑:肖 莉
助理编辑:齐梦涵
版式设计:吴 迪

广西师范大学出版社出版发行
(广西桂林市五里店路9号 邮政编码:541004
网址:http://www.bbtpress.com)
出版人:张艺兵
全国新华书店经销
销售热线:021-65200318 021-31260822-898
恒美印务(广州)有限公司印刷
(广州市南沙区环市大道南路334号 邮政编码:511458)
开本:635mm×965mm 1/8
印张:34 字数:30千字
2018年5月第1版 2018年5月第1次印刷
定价:288.00元

目录

前言

建成项目

在建及未来项目

改建项目

GOETTSCH PARTNERS

大胆的构想与对细节的执着追求是该公司自创立以来一直坚持的特色。

——评审意见

美国建筑师协会芝加哥分会2013年度公司奖

公司介绍

概况

盖特可事务所是一家具有全球视野的建筑事务所。本公司总部位于芝加哥，另在上海和阿布扎比设有分公司，我们采用通过多年实践累积的经验和对探索与创新的热情汇聚而成的方法进行设计。我们建成并投入使用的项目在类型和规模上各有不同，范围横跨五大洲。

目标

盖特可事务所力求将客户的要求和野心转化为有意义的结构，努力打造出能够对周围环境乃至更广泛的社区产生积极影响的建筑。我们深思熟虑地考量设计的价值和技术的特色，以期为项目带来可量化的价值，同时也为我们的客户提供高水平的服务。

方法

我们在设计前不会预先设定好审美风格，而是先从对各个概念的评估开始，从零启动一个设计过程，我们主要从以下几个角度进行评估：内在价值、可施工性、对环境的影响。我们同每一位客户通力合作，力求在我们设计的项目中实现他们的目标与愿景。我们与世界级的工程师及专业技术顾问密切合作，以确保每个项目中的特殊要求都能够得到满足。

前景

我们擅长设计强调透明度和空间活力的大胆而清晰的建筑，也一直希望借助这种建筑提高建筑周边的环境品质。我们的项目拥有一致的视觉语言，这有助于促进高技术的解决方案，提供最佳的使用性能。

传承

盖特可事务所的领导班子是不断发展着的，其历史可以追溯到由密斯·凡·德·罗1938年在芝加哥开创的崭新实践方法。詹姆斯·盖特可（James Goettsch）于1992年加入到对这种实践的探索中来，他在这个过程中培育出了一种可传承的文化，这种文化不断激励我们寻找属于我们自己的、适合21世纪的永恒物质性和清晰的建筑表达。

领袖人物

合伙人

詹姆斯·盖特可是盖特可事务所的董事长兼首席执行官，他同时也是公司的设计总监，负责为公司培养设计人才，管控公司的设计方向与设计质量。

JAMES GOETTSCH, FAIA

詹姆斯·郑 (James Zheng) 是盖特可事务所的总经理，负责全公司的总体业务、管理策略，以及日常经营。他参与的项目横跨美国及海外，他从这些工作中得到的丰富经验有助于指导盖特可事务所在全球的扩张。

JAMES ZHENG, AIA, LEED AP

乔·多利纳 (Joe Dolinar) 是管理合伙人，他参与了多个规模巨大且极具挑战的任务，其中包括位于美国的若干总部大楼、大型写字楼及军人球场的改建项目。

JOSEPH DOLINAR, AIA

迈克尔·考夫曼 (Michael Kaufman) 是项目管理合伙人，负责领导和协调大型设计团队。他参与过的项目类型多样，从国际酒店和办公大厦到学术和机构设施，再到创新改建工程，等等。

MICHAEL F. KAUFMAN, AIA, LEED AP

拉里·韦尔登 (Larry Weldon) 是技术合伙人，他负责领导盖特可事务所的围护，这是一个公司内部的专门小组，主管外部建筑围护结构和系统的各个方面。他利用自己在这个领域内的技术知识，确保了设计图纸及其在整个设计流程中的完整性，也使它们的清晰性和一致性得到了保障。

LAWRENCE C. WELDON, AIA

主管

VLADIMIR ANDREJEVIC, AIA

CLAIRE YIN CHEN, CLASS 1 REG. ARCHITECT

PAUL DE SANTIS, LEED AP

LEONARD KOROSKI, FAIA, LEED AP

PATRICK LOUGHRAN, FAIA, PE, LEED AP

JOACHIM SCHUESSLER

SCOTT SEYER, AIA, LEED AP

TRAVIS SOBERG, AIA, LEED AP

ELIAS VAVAROUTSOS, AIA

STAFF
员工

CHICAGO, SHANGHAI, ABU DHABI: Raj Achan, Vladimir Andrejevic, Richard Barrett, Scott Brackney, Yangmei Cai, John Campbell, Zhexiong Chai, Randall Chapple, Claire Yin Chen, Wenfei Chen, Kenny Chou, Paul De Santis, Joseph Dolinar, Grace Faoro, Andrew Fox, Daniel Fragakis, Chunjiang Fu, James Goettsch, Jay Grodecki, Matthew Hall, Erik Harris, Aaron Haslerig, Silin He, Yizhou He, Lisa K. Hill, Jonathan Hoffman, Nathaniel Hollister, James C. Horton Jr., Jinwoo Jang, Xuezi Jia, Philip Johnsson, Matthew Johnstone, Alex Kang, Michael F. Kaufman, Jason Kocher, Leonard Koroski, Alesia D. LaCaze, Matthew C. Larson, Jin-wook Lee, Jin Li, Xiaohai Li, David J. Lillie, Daniel J. Lipetzky, Ian Liu, Patrick Loughran, Yue Lu, Zhengting Luo, Xiao Ma, Kathleen Maciejko, Marcus McLin, Brian C. Miller, Maria Miller,

Cecilia Min, Frank Mraz, William B. Netter, Laura Niekamp, Brad Novak, Al Ochsner, Stephanie Pelzer, Nickolas Popoutsis, Susan G. Pratt, Daniel M. Ramos, Logan Reed, Hiroshi Sango, Heather Schmitz, Joachim Schuessler, Joseph Schultz, Scott Seyer, Ye Sheng, Denise Sims, Travis Soberg, Peter Stutz, Feifei Sun, Afaq Syed, Kirk Tracy, Genevieve Trindade, Elias Vavaroutsos, Melissa Venoy, Michael Wagenbach, Hai Alex Wang, Wei Wang, Adam Weissert, Lawrence C. Weldon, Charles W. Wittleder, Aliyah Wu, Di Wu, Liuqingqing Yang, Difei Yao, Seungpum Yoo, Ni Zhang, James Zheng, Gang Zhou, Lillian Zong, Ed Zimmerman

Left to right: Paul De Santis, Elias Vavaroutsos, Scott Seyer, Joachim Schuessler, Leonard Koroski, James Cramer, James Goettsch

从左到右：保罗·德·桑蒂斯、伊莱亚斯·维瓦鲁特索斯、斯科特·塞耶尔、约阿希姆·舒斯勒、莱纳德·科罗斯基、詹姆斯·克莱默、詹姆斯·盖特可

与盖特可事务所对话

美国建筑师协会荣誉会员，詹姆斯·P. 克莱默

难得有机会坐下来与盖特可事务所的设计领导者们一起，进行这样一次有关公司前景的非正式谈话。我们几人讨论的议题很广泛，既包括盖特可事务所的现状、成就，也包括公司未来的走向。

我之前曾在美国建筑师协会担任首席执行官，在未来设计协会担任名誉主席，与不少建筑公司有过合作，也写过多篇与建筑和设计专业问题有关的文章，公司的领导们因此邀请我来主持这次谈话，我的观点曾帮助盖特可事务所阐明对他们来说很重要的问题，使他们专注于这些问题，例如不断变化的技术实践、大胆打破现状，使他们在每个项目上无论是对内还是对外都能创建出一个相互信任的环境。

下面就是我们这次全天谈话中的一些重点。

根植芝加哥，放眼全世界

詹姆斯·P. 克莱默（JPC）：盖特可事务所的作品已经在媒体和房地产出版物上引起了非常多的关注。今天早上去你们的图书室时，我路过工作室看见了里面正在进行的最新项目的模型。它们不仅展示出了公司在设计上的野心，更重要的是，它们展现出了以人为本和不断发展变化的设计方法。我们就从是什么定义了今天的盖特可事务所这个问题开始我们今天的谈话吧。

詹姆斯·盖特可（JG）：我们十分看重我们是芝加哥的建筑事务所这一现实。芝加哥的建筑历史悠久丰富，能为我们提供许多宝贵的经验教训。早在19世纪八九十年代，芝加哥的建筑师们就开始探索深层地基的建筑方法，他们还利用结构框架令大面积使用玻璃成为可能。他们脱离了因历史渊源产生的装饰风格，率先在建筑物里安装电梯、空调等装置。这些都是"现代建筑"的种子，这一时期也因为著名的"第一芝加哥建筑学派"而举世闻名。

在20世纪50和60年代，密斯·凡·德·罗依据自己在伊利诺理工大学的教学和他自己的建筑作品，开创了第二芝加哥学派。这一时期，无论在理论上还是实践上，建筑师们都开始强调在结构中蕴含艺术的可能性。密斯为全世界的建筑事务所带来了启迪，并进一步巩固了芝加哥的设计行业领导者的地位。

考虑到我们的历史，我们就会有一种开拓进取、力求创新、提升价值的心态。我们作为一家芝加哥建筑设计公司，却能把业务范围拓展至全球，这让我感到很自豪，我们的建筑方式虽然是根植于芝加哥建筑传统之中的，但是我们也同样能够追上快速变化的设计和施工技术。

保罗·德·桑蒂斯（PD）：我们今天在盖特可事务所创造的建筑，在本质上是超越时间的，而不仅仅是新潮及时尚的。密斯留下的丰富精神遗产，直到现在仍然不断在设计哲学和实践过程方面启迪我们，但是我们本身也仍需每天努力推动我们当前的项目，把我们的设计方案打磨得更加智能、更具启发性，让它们成为我们这个时代的标志。

莱纳德·科罗斯基（LK）：芝加哥建筑在设计和技术创新方面都有着丰富的历史。我们公司的作品既欢迎也强调了这些历史，尤其是在保护、改造和改造性再利用等领域。从这一传统出发，我们将继续打磨我们的"芝加哥方法"，并将其融入到我们在全球的工作中去。

伊莱亚斯·维瓦鲁特索斯（LV）：我们继承的建筑遗产有其固有的严谨和真实的需求，也有创新的想法，这些想法需要以大胆清晰的方式来表达。我们希望在技术创新和创造具有特定感觉的场所之间找到一种平衡，希望我们的建筑既简单又优雅，同时在语境和文化中具有意义和共鸣。

JG：我想补充的是，作为一家总部设在芝加哥的公司，我们工作务实、严谨，坚持寻找改进的方法。我们重视设计与施工之间的关系，知道两者之间的融洽关系能够提升我们建筑作品的质量。

深思熟虑的方法，永恒的解决方案

JPC：盖特可事务所是如何改变、进化和变得更加多元化的呢？

斯科特·塞耶尔（SS）：我们着手的每一个项都根植于项目所在的地点和环境，例如，我们在旧金山面临的挑战就与在芝加哥、迪拜或上海的不同。市场趋势也随着不同地点的地理、人口、技术条件及不同的文化背景而变化着。在这样一个加速变革的时期，我们的目标就是在每一个我们涉足的市场中为我们的建筑增添价值。本着这一精神，我们一直在寻找能够挑战当今惯例的潜在突破口。

LK：从我们位于芝加哥的办公室向外看，我最喜欢的风景之一便是瑞士银行大厦，它的底部装有美国最早一批缆索承重的玻璃墙。而它身后则是坐落在街对面的历史悠久的布杂风格城市剧院，我们曾沿着那座建筑的门廊修复了它足足有9米高的装饰性铸铁正立面。这两座建筑在刚建成时都运用了当时的前沿技术，建筑师们在这些项目中提供了独特的设计解决方案，并开创了一种充满活力的设计方式，每座建筑也都很好地对自己所处的时代做出了表达——这些都是超越时间的解决方案。这种方法也为我们从一个全新层面为建筑添加附加价值做好了准备。我们有信心抓住这个时代的脉搏，但是我们也必须继续学习。

LV：我们在综合整体规划方面的工作能力也有大幅度的提高。在大的开发项目上，我们一般会先规划设定一个整体基调，然后再着手进行具体的建筑设计。我们还承接非传统办公大楼，例如公司总部大厦、创业公司的办公楼和为联合办公所建的大楼。

PD：我想强调的是，我们虽然有创造出具有雕塑之美的建筑，或是实现某些激进的形式的欲望，但是我们不是简单地被这些欲望所驱使的。我们在开展一个新项目前，总会确保我们的设计者没有先入为主的审美取向，再根据项目的环境影响、内在价值和经济足迹来评估摆在我们面前的选择。这个过程是必须的，因为我们的目标是在世界各地建设各种各样的大楼。

JG：我们不想自夸，也不想与其他公司进行比较。事实上，我们在许多方面都和其他设计公司有相似之处。对于我来说，问题不在于我们有多好，而是我们想成为一间多好的公司。

LV：我们承接的许多项目都是讲求务实的，它们以自身的某些意义，使自己与其他建筑区别开来。我们的建筑不是为了压制其所处地点与所容纳的人而建的，它们从来就不是浮华夸张的。我们所设计的大量作品都包含某种一致性，例如，它们的质量令人满意，随着时间的推移，我们希望人们从初次见到这些建筑所产生的震撼感中走出来，慢慢熟悉它们，并对这些建筑有更深的理解。正因如此，建筑的使用寿命和耐用程度对我们来说非常重要。

SS：我们的核心设计理念是将用户体验纳入考量。我们不单要考虑环境，还需要了解文化以及文化是如何影响项目的实质的。且不论文化影响，我们相信，当人们走进一座盖特可事务所设计建造的大楼时，他们会直观地感受到建筑的质量很好，设计很适宜。

LK：我想强调的是我们现在和过去的建筑工作是相辅相成的——相互承认和接受设计及技术解决方案，并用务实的方式来支持城市环境文脉。

可持续发展、创新和持久价值

JPC：可持续发展的下一步是什么？你们会引领创新吗？

约阿希姆·舒斯勒（JS）：我们的可持续性标准不会因地区不同而区别对待，因为这是我们建筑实践的一个核心价值。当然，最让我们激动的事情莫过于能够达到最高可持续性评级，或某个项目收到奖项的肯定。

JG：我们拥有成熟的客户，他们是我们努力在可持续性设计中达到更高标准的合作者。然而，我们还可以做得更多。我们需要更好地了解建筑如何利用能源，怎样才能降低运营成本，以及我们还能做哪些事情来改善我们建筑物的内部环境。关键在于我们有创造更高水平的可持续性设计的强烈愿望，它也是我们公司的思维模式之一。

PD：我们已在本地和海外建造了一些铂金奖绿色建筑和金级认证的建筑。这是评价我们在可持续性方面取得的成就的合适指标吗？恐怕不是。我们有责任设计出更健康，能为建筑的用户和更广大的社区带来积极影响的建筑，我们需要继续挑战自己以及为我们的客户做更多的事情。

SS：没错，我还要强调一下作为企业文化的一部分，我们应该努力在可持续性方面进行创新尝试。我们希望一旦有机会就把这种实践推进到下一个层次。我们聘请了能够丰富我们团队的专家顾问。总之，这种方法有助于项目向更高水平的可持续性设计迈进。我们专注于定位、体量、材质等，为用户能够获得最舒适的体验而创造可持续性建筑。

JS：与美洲和亚洲相比，欧洲的能源标准有很大的不同。例如，我们有一个位于华沙的商业高层项目，当地的能源法规要求建筑的围护结构必须设计为三层绝缘玻璃，这在美国是很少会被考虑到的。虽然如此等级的可持续性承诺并不是在哪里都会被要求的，但是我相信对可持续性的理解在今后会继续发展。这个问题有很多层面，例如可持续性意味着你需要一个高效且空间灵活的楼面，以及开放式的空间，因为这种布局可以提高工作效率。未来几年会有更多这种布局的建筑落成。这个领域为我们提供了一个战略机遇，我希望我们能为推动这一事业做出贡献。

PD：另一方面，玻璃、照明、建筑外壳与其他材料和系统为客户和用户们带来直接的可持续效益和其他方面的好处。当我审视我们公司和项目现如今的复杂性时，我看到了我们在材料方面的专业知识，并看到了这种优势为我们带来的提升建筑价值的能力。

JPC：你们的工作以创新方法著称，尤其是为解决客户的问题时提出的创造性解决方法。请给我们详细讲讲盖特可事务所在未来还会给我们带来哪些期待。

LV：创新是由主动性和必然性驱动的。在我们的项目进行的过程中，建筑功能和规范要求在设计阶段经常有变化，所以需要考虑的对象和优先事项有时是不固定的。我们必须提出正确的问题，然后在应对这些问题时保持灵活性和创造性，以期继续为我们的客户提供有价值的服务。

JS：我们应该强调的是，我们在建筑中实现创新的过程并不是线性的。它们是逐渐积累产生的，它们随着世界各地的动态而变化。当我们说我们通过创新创造价值时，我们指的是我们不仅具有专业技能，而且具有创造性的悟性，同时明白怎样运用它们。

PD：现在是加速发展的时代，变化的速度比以往任何时候都快。技术对我们的生活和工作产生的影响是深远的。我们现在可以使用复杂的协作工具来设计和评估我们的项目，而这些工具在两年前还不存在。创新不仅是我们实践的一个目标，它也是我们对每个项目的期待。

全球化经验、专长和对表达的热情

JPC：人们在走进盖特可事务所时，会感觉到有什么东西是不一样的，它是什么呢？它是如何在战略层面上使这家公司变得与众不同的呢？

JG：这个行业现在已经发生了变化，我们也已经适应了这种变化。当我还年轻，刚成为一名建筑师时，新的工作机遇对于大多数公司来说常常来自于社会、政治和商业的联系。如今，新的商业机会已经建立在经验和专业知识的基础之上，人们对专业知识的依赖比以往任何时候都要强烈。幸运的是，我们已经把我们的公司打造成了这种模式，这使我们可以充分发挥专业的力量。

我们已经在全球工作20余年，已建成的项目数量十分可观。已经竣工了的建筑项目超过120个，在亚洲、中东、欧洲、美洲还有超过20个正在建设的项目，另外还有15个处于设计阶段。虽然我们不是一家大公司，但我们在商业写字楼、酒店和多功能建筑方面的经验和专业知识使我们能够与世界领先的大型建筑设计公司竞争。我们的施工质量也在世界范围内获得一致好评。

LV：吉姆关于专业知识的说法在亚洲表现得尤为明显。我们在中国的业务一直很繁忙，我们成功地吸引到了客户并维持住他们对我们的信任，这些客户有房地产商、建筑商，除此之外还有各种与建筑商、地产商并列不到一起。尽管目前经济增长放缓，但这些客户依然保持着经济实力，能够继续购买和开发土地。他们把建筑看作建立多元业务的手段，他们了解我们的经验和声誉，知道我们是一家能够创造出杰出作品，关心客户的商业成功的公司。

JS：在中国的竞争压力越来越大，这不奇怪。我们希望能因我们的设计质量与服务质量而成为客户的首选。

SS：我们尝试去理解客户的价值观，这对创建一个成功的项目十分有益。虽然我们的大多数项目都是在与老客户合作，但我们仍然在不断的努力尝试与新客户建立合作关系。我们都清楚房地产行业面临的经济挑战，我们希望我们的客户能够从我们对这个行业的深刻了解中受益。

JPC：看来公司的愿景是显而易见的，我们能感受到这其中的使命感、方向感和目的性。能不能再和我们分享一些你们的想法呢？

LK：公司的愿景是在不断变化着的，因为我们在高层建筑、机构建筑和重大老建筑保护项目方面均有涉猎，工作具有多样性。我们在美国和全球的工作经验允许我们“看得更多”，有更多的机会帮助美化街道景观、提升每一座建筑的价值。特别是在老建筑保护与改建这一领域，我们有着广阔的未来，如果把这一领域的工作也包括进来，那么我们作为一家公司的经验也就会变得更加成熟了。

JS：很多建筑公司在第一眼看上去的时候是非常相似的，而我们的愿景之一就是继续保持使我们与其他公司区分开的独特个性。虽然每个项目都各有不同，但我们依然要保证为客户提供世界级的服务和独特性。每个市场都有差异性，我们的责任是保持灵活多变和消息灵通，在重要的决策上为我们的客户提供更加深入的见解。

PD：我们对未来的憧憬也集中在对创造一种质疑文化——不断提出问题——的渴望上。我们怎样才能做得更好？我们希望我们的工作室成为高水准的材料和表现的资讯中心。我很高兴这里的同事和我有着同样的创新热情。

JG：我们在乎我们的工作。我们每个项目的负责人都是亲自动手参与实践的，他们会在项目的各个阶段从哲学、智能和科技等层面全力满足我们客户的需求。我们的公司财务状况良好，但是我们的项目经理从不把时间都用在保护我们公司的利润上。我们的主要目标是实现客户的目标，只要这样做，我们的财务状况必定会一直保持这样的理想状态。

LV：我们想要承接最好的项目，占据最好的地点，与优秀的客户合作。我们希望交付有意义的作品，它们应该是从区位特点、文化共鸣和使用体验等工作理念中产生的解决方案。

创业精神，技术优势

JPC：这次谈话体现出和当今关键问题相关的深刻见解，以及盖特可事务所的秩序与方向。还有什么你们想要告诉我们的其他价值观或理解吗？

PD：我们还必须强调我们员工的力量。我们积极寻找多元化、有才干、有活力的个人，他们理解我们公司的价值，也愿意帮助这家公司拓展更多的新项目。年轻的成员从公司领导者们具备的丰富经验中不断学习，茁壮成长，为公司做出更大贡献。他们对新技术、新软件及建模方法的了解，使他们成为我们作为一家设计公司要想不断发展不可或缺的重要组成部分。

LV：我们想要提出最好的想法——不仅是好的，还必须是最好的。这些想法可以来自团队中的任何人，无论他的经验或是专业知识如何。我们能认识到整体合作在实现成功中的价值，也希望找到火花。我们明白年轻而充满活力的员工知道他们能对总体设计产生影响时所产生的那种自豪的感受，他们会对其为之努力工作的项目和公司的氛围产生更大的认同感。

SS：我们的团队一直为每个项目寻找创新方案。无论是在建筑的功能、技术、体量还是物质方面，我们总会挑战自己。随着公司的发展和年轻员工经验的积累，我们势必能够更加积极地合作，寻找到最棒的想法。

PD：公司内部的竞争精神也在增强。我们提倡这种精神，这种精神越高涨，我们的项目就会越丰富多彩。我们以新的方式来创造、评估和发展。我们想以创业精神和包容的态度面对新的机遇。创业精神是从吉姆和詹姆斯身上开始的，我们不仅仅是想保持公司现有的标准，我们希望能以热情和积极的乐观精神面向未来。

JG：毫无疑问，在过去，刚入行的年轻建筑师在竞争力上是落后的，但是如今，他们才是掌握最新技术和软件知识的人，在这些刚出校门的年轻建筑师身上，我们会发现真正的价值。不是说过去的毕业生没有现在的毕业生愿意在各种实践中做出贡献，之所以会有现在这样的情况，是因为我们这个行业的基础是知识，而学习知识需要时间。现在的毕业生对最新的建筑信息模型、参数化设计和最新的可视化软件了解充分，年轻的建筑师可以从一开始就赢得自己的地位。

SS：我们始终希望提升我们的软件平台，寻找新的数字工具来加强我们的工作。比如，我们的办公室现在使用虚拟现实软件查看项目的内部结构，这可以允许我们从新的角度看待某些设计思路。无论是使用新软件还是交互式、协作式的智能演示版，我们都积极采用新手段推进项目设计，这对项目是大有益处的。

JPC：你们从哪里招募这些有才干的人呢？

PD：我们现在已经瞄准了中西部地区和东海岸的几所重点学校。这些学校目前的课程与我们现在的设计方式非常吻合。虽说如此，我们同时也从全国及全世界的其他学校征集作品，从中聘请优秀的申请者加入到我们中来，提升我们的实践能力。

SS：我们公司也在寻求与高等教育机构直接对话的机会。我们访问了许多我们此前招聘过员工的大学，去了解他们的课程设置，学生能够接触到哪些设备等信息。这确实帮助我们了解了我们许多年轻设计师的背景，我们想要寻找的是欢迎并享受和我们做同样事情的课程项目。

JG：让我觉得有趣的一点是，也许我们公司现在比以往任何时候都适应与高等教育的接轨。新科技正在改变我们开发、建设、评估建筑物的方式，而每一个新的毕业生都能带来新的技能。

LK：老建筑保护领域的很多技术也是新的。我们在工作的过程中会采用新技术，例如用扫描和嵌入的方式把过去的设计加入到我们现在的工作之中，激光三维扫描空间既可以使我们看到一个点云中的退化或结构位移，也可以帮助我们更改建模或修缮改建。我们使用热扫描仪查看和追踪墙面状况和水的渗入情况。在生产中，我们创建装饰照明的3D模型，使复原或复制原型更加快速便捷。我们以前要花12周来进行的工作现在只需要两周。

公司还以新的方式在现有的建筑物上推行建筑信息模型的使用。我们可以采用新手段使建造实物模型的时间减少一半，实现快速建模。

JPC: 你们认为技术还将会怎样挑战和改变我们现有的工作方式?

PD: 首先是生产力。每位全职员工都能为公司做出更大的贡献，且效率更高。但是，更快的交货期和更高的质量是我们的客户如今的要求标准。因此，新技术和扩大的设计服务会继续使公司的员工数量不断增加。

JG: 不久之前，施工文件还要由几个独立的小组在各自的学科中编写。而现在，我们都在使用建筑信息模型，我们或多或少都在使用同一个可以整合与协调复杂组件的平台。

标志性建筑，城市的焦点

JPC: 以盖特可事务所的标准，什么样的建筑才是一个优秀的城市地标?

SS: 建筑应该使用营造一种欢迎人们到来的气氛。优秀的建筑物可以使城市变成更好的地方。我们的建筑是催化剂，目的是对其用户与其周围的环境产生积极的影响。

JG: 在过去的15到20年里，“标志性”这个词语在建筑学讨论中已经变得司空见惯。我们在听到“标志性建筑”时会不由得畏缩。因此我们更喜欢使用“地标建筑”这个词。

我们的一位客户喜欢谈论他们对建造一座A级建筑的愿望。术语是一个表达明确价值的声明，而A+级建筑的定义非常简单：或者达到最高的租赁率，或者在当时与其他建筑相比，每平方米的售价最高。开发商专注于建筑作为投资工具的功能，而我们希望看到我们的客户能最大限度地实现他们的投资目标。每一个项目都有一定的潜力，而感知这种潜力则是一门艺术。如果你把目标定的太高，可能会耗尽项目的资源，这会导致预算超支；如果你的目标太低，则可能无法最大限度地利用项目的潜在资源。而要使效率和经济效益最大化，我们的客户相信他们需要通过多种手段来为项目增加价值，例如提供模范性的工作环境、最高水准的建筑体系、优秀的设施，以及能为周围社区增加价值的街景与天际线。换句话说，就是建造一座“地标建筑”。

JS: 虽然已经被提到过，不过我还想再重申一次，我们关注城市，但是也关注人。我们在老地方或新区域都能看到价值。虽然我

们对城市复兴感兴趣，不过我们也喜欢为共同的利益而变革。现在对于实践工作，共同塑造我们的环境而言是一个有趣的时代。可持续发展是我们的首要任务之一，我们不是以革命性的方式，而是以渐进的方式来实现它。我们的目标是让居民在城市环境中感受到幸福。

SS：我们力求以多种方式对项目的直接环境和周围环境产生积极影响。我们希望在街面和天际线两方面都创造出视觉趣味性。我们想要超越期待，创造出有趣的、吸引人的、功能性强的建筑来改善它们所处的环境。

PD：20年前，我就被这家公司吸引住了。设计与建造工艺的结合为房地产市场树立了新的标准。这些传承下来的历史建筑至今依然保持着活跃，而我们目前正在建造的项目将在市场上设立新的标准。我们的建筑物对环境和使用者做出的回答比以往任何时候都先进。

JG：我们是由一家以设计郊区公司总部大楼著称的公司发展而来，因此，我们目前的作品集，那些最能与盖特可事务所联系到一起的作品，都是从大约20年前开始积累的。

我们的目标从一开始就是设计“地标建筑”，我们在大型商业高层建筑上花了很多力气，特别是写字楼、酒店和多功能复合建筑物。因为我们是从零开始积累我们的作品与客户，所以我们总会意识到每一个项目的成功都是我们下一个项目的基础。我们从不认为获得哪个项目是理所当然的。

当我们说这是一个成功的项目时，我们指的是这座建筑可以匹配或超越客户的预期功能目标，在某种程度上，我们已经在技术方面挑战了建筑的极限，整个街区也认为这是一个具有“地标”意义的建筑，这表明我们的建筑为街区景观与天际线增添了价值。

当我们在寻找新机遇时，我们也要坚守我们的创新品格，持续保持适当的克制、谦卑和勇气，这不是为了我们自己，而是为了大家共同的利益。

建成项目
CURRENT

河畔北150号
150 NORTH RIVERSIDE

Chicago, Illinois, USA

The 150 North Riverside site is located prominently at the confluence of the three branches of the Chicago River and less than one block from one of Chicago's busiest commuter train stations. With exposed railroad tracks on the west side of the site and the city requirement for a riverwalk on the east side, the remaining area on which to build was considered impossibly narrow, and the site sat undeveloped for decades. As most of the viable office sites were built over time, the developer decided that an unconventional approach was required for this prime parcel.

Utilizing a unique core-supported structure with a very small footprint at grade, the design resolves the site challenges and provides a 54-story Class A office tower with efficient, 45-foot column-free floor plates. Tenants and visitors enter through a dramatic, 90-foot-high lobby enclosed by a glass-fin wall hung from the structure above. The lobby features a 150-foot-long curated multimedia wall that showcases the work of local and other digital artists across 89 LED blades. The site-specific installation provides a focal point for the space while also addressing the transition between the opaque wall over the parking deck and the start of the glass-fin wall.

Vertical mullions on the exterior take cues from the river and undulate along the building's wide east and west façades to help activate them with an ever-changing pattern of light and shadow. The narrow north and south faces are divided into three vertical planes that accentuate the slenderness of the tower, with the center plane recessed to create additional corner offices. The very condensed lobby and elevator cores allow the majority of the two-acre site—more than 75 percent—to be a landscaped public park with pedestrian pathways overlooking the river. Building amenity spaces include a restaurant, bar, fitness center, and conference center—all with water views.

Designed for LEED Gold certification, the project addresses several sustainable initiatives as it connects and revitalizes a critical downtown parcel, adding a distinct office tower and active new urban spaces for the city.

This is a gutsy building, one that rises to the challenge of its site.

—Blair Kamin, Architecture Critic, *Chicago Tribune*
"Tower Rises to the Challenge," April 23, 2017

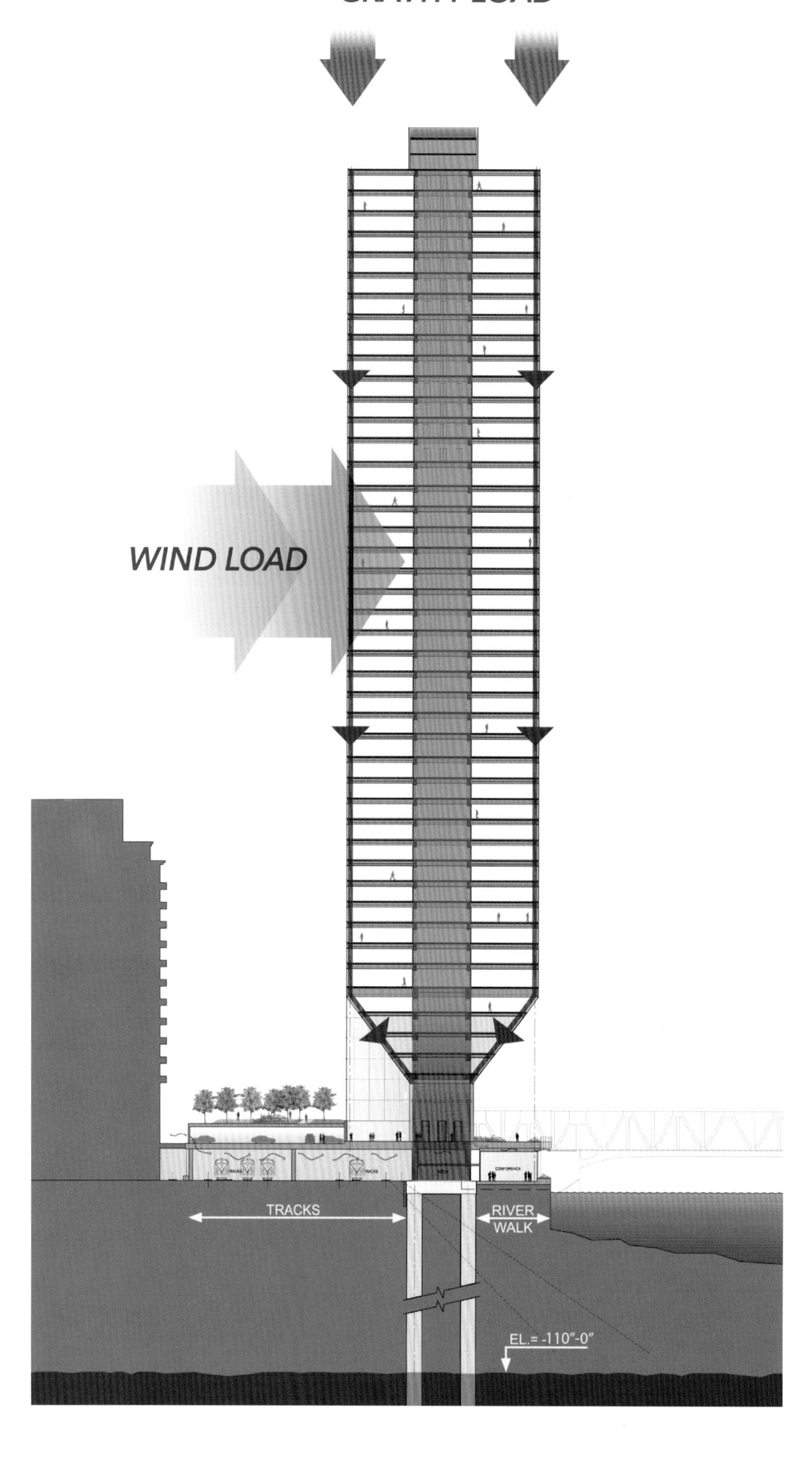
GRAVITY LOAD
WIND LOAD
TRACKS
RIVER
WALK
EL.= -110"-0"

RANDOLPH STREET

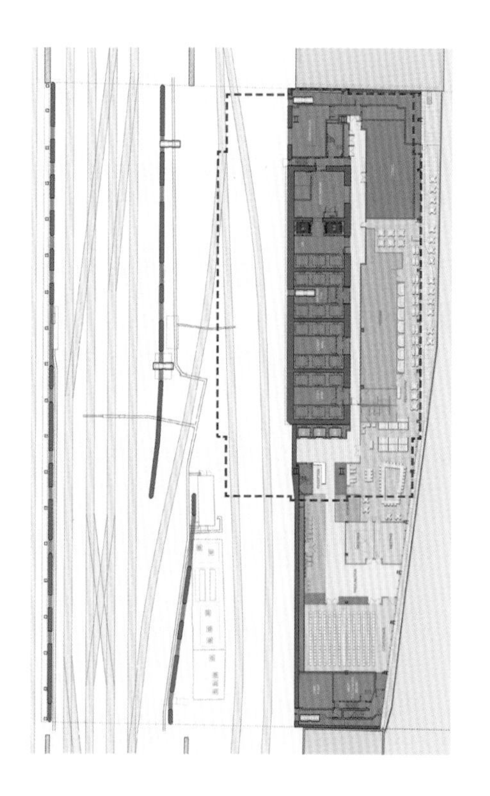

TRACK-LEVEL PLAN

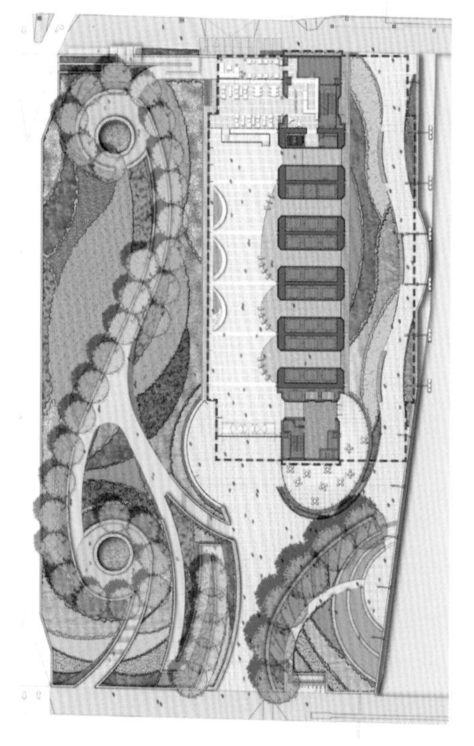

LOBBY- AND PLAZA-LEVEL PLAN

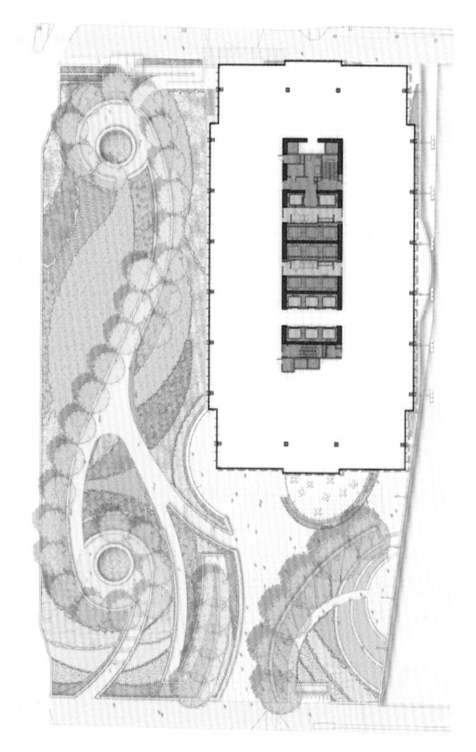

TYPICAL FLOOR PLAN

苏黎世保险公司北美总部大楼

ZURICH NORTH AMERICA HEADQUARTERS

Schaumburg, Illinois, USA

Located on a 40-acre expressway site in suburban Chicago, the North American headquarters of Swiss-based Zurich Insurance Group is designed to reflect the company's global reach and world-class stature. The formal architectural resolution strives to represent both strength and stability, which are core values of the Zurich business model. Composed of three primary "bars" that are offset and stacked, the arrangement creates unique spaces for collaboration, opens views of the surrounding landscape, optimizes solar orientation for amenities, and provides programmatic flexibility not found in typical center-core office buildings. The top "bar" of the complex soars 11 stories and cantilevers toward downtown Chicago, providing visual identity along the interstate while projecting the strength and future focus of the company.

Certified LEED Platinum, the 783,800-square-foot complex reinforces Zurich's commitment to environmental stewardship. A network of horizontal sunshades clads the perimeter of the complex, with the sunshades varying in depth depending on orientation, while floor-to-ceiling glass offers extensive natural light for the shallow office plates. A soaring three-story double wall faces south toward the multi-level plazas, showcasing an architecture that responds to the changing Chicago climate. In the end, a timeless material palette married to a bold, clear form creates a unique identity for the Zurich headquarters that embodies the core values of the company.

Its synthesis of structural expression, sustainability and architectural showmanship makes for a powerful display...

–Blair Kamin, Architecture Critic, *Chicago Tribune*
"Suburban Office Complex Breaks with Convention,"
September 25, 2016

ZURICH
ZURICH

Z
ZURICH

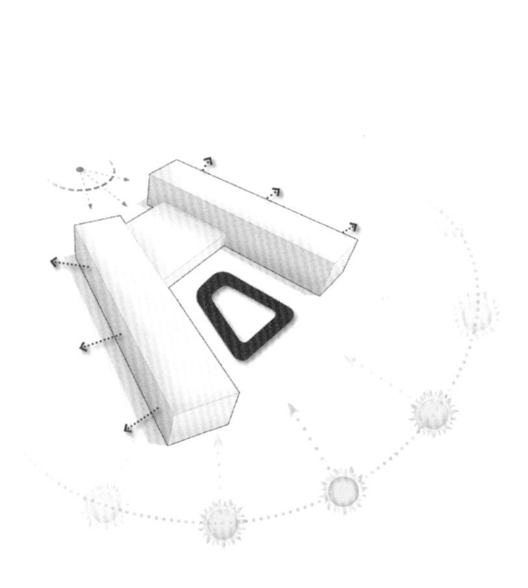

INTERIOR COURTYARD OPENS UP TO THE SOUTH, RECEIVING CONTINUOUS DAYLIGHT

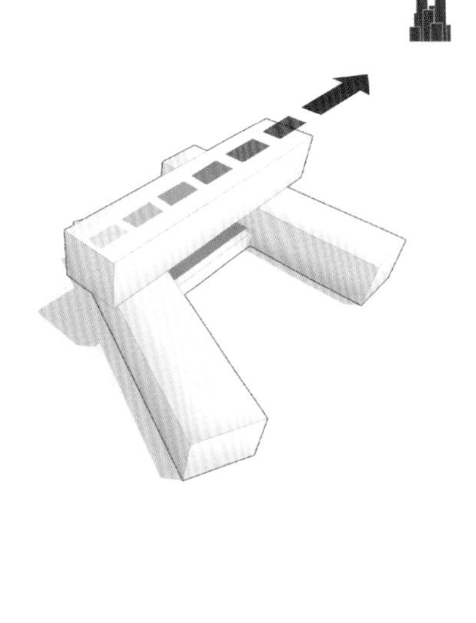

THE TOP BAR OF THE COMPLEX CANTILEVERS TOWARD DOWNTOWN CHICAGO

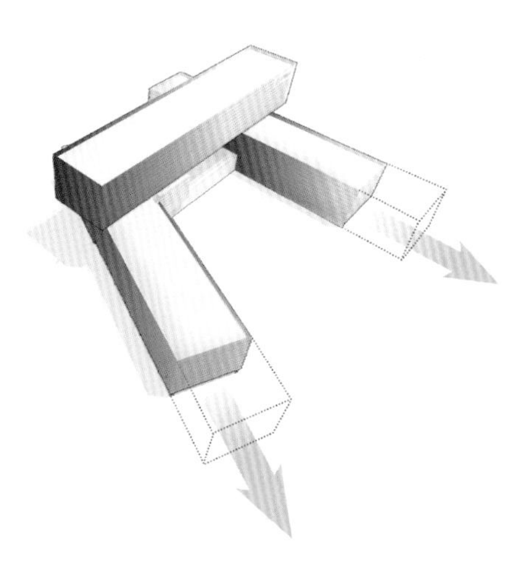

LOWER BARS HAVE POTENTIAL FOR EXPANSION AS NEEDED

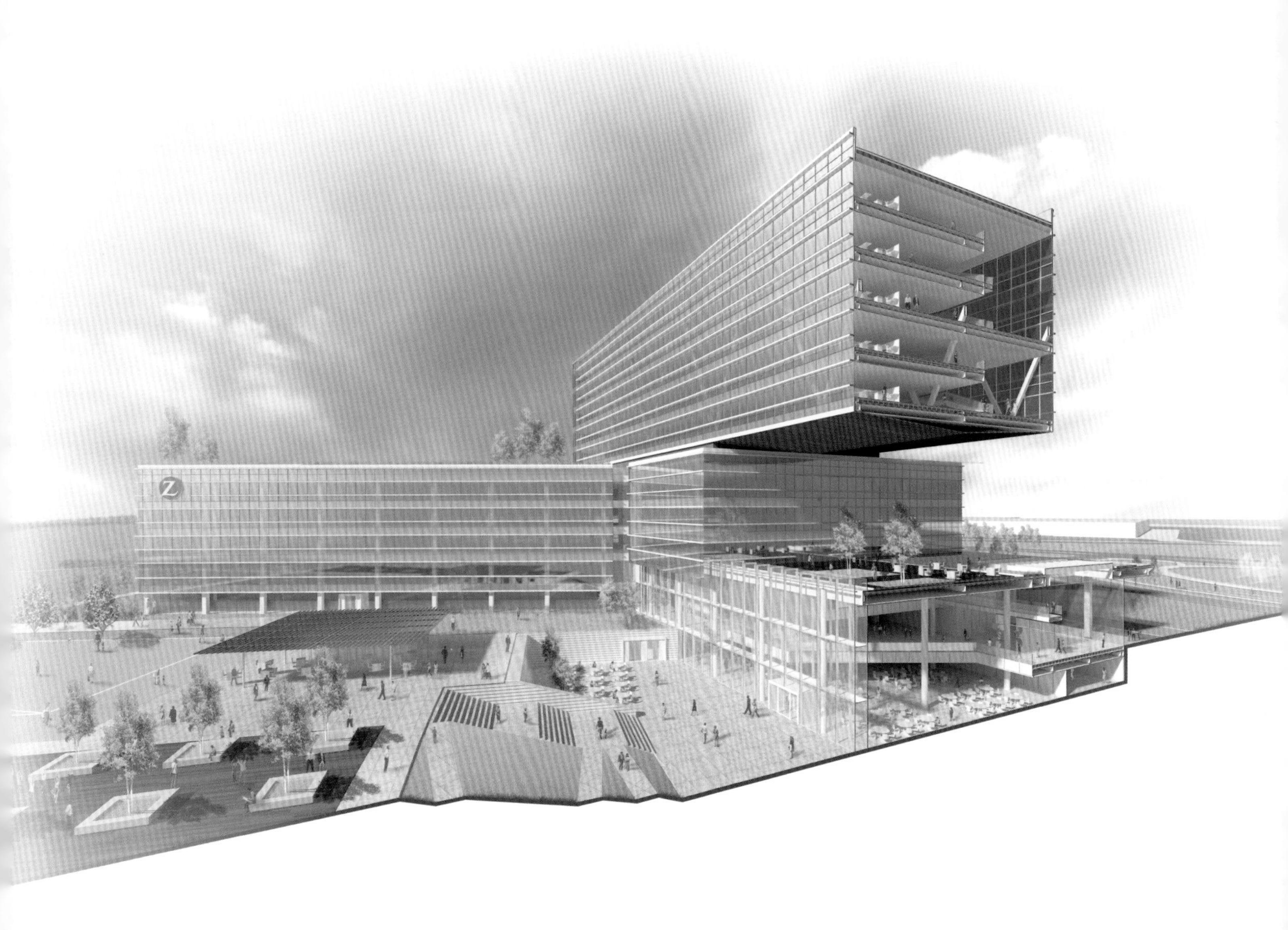

ZURICH

ZURICH

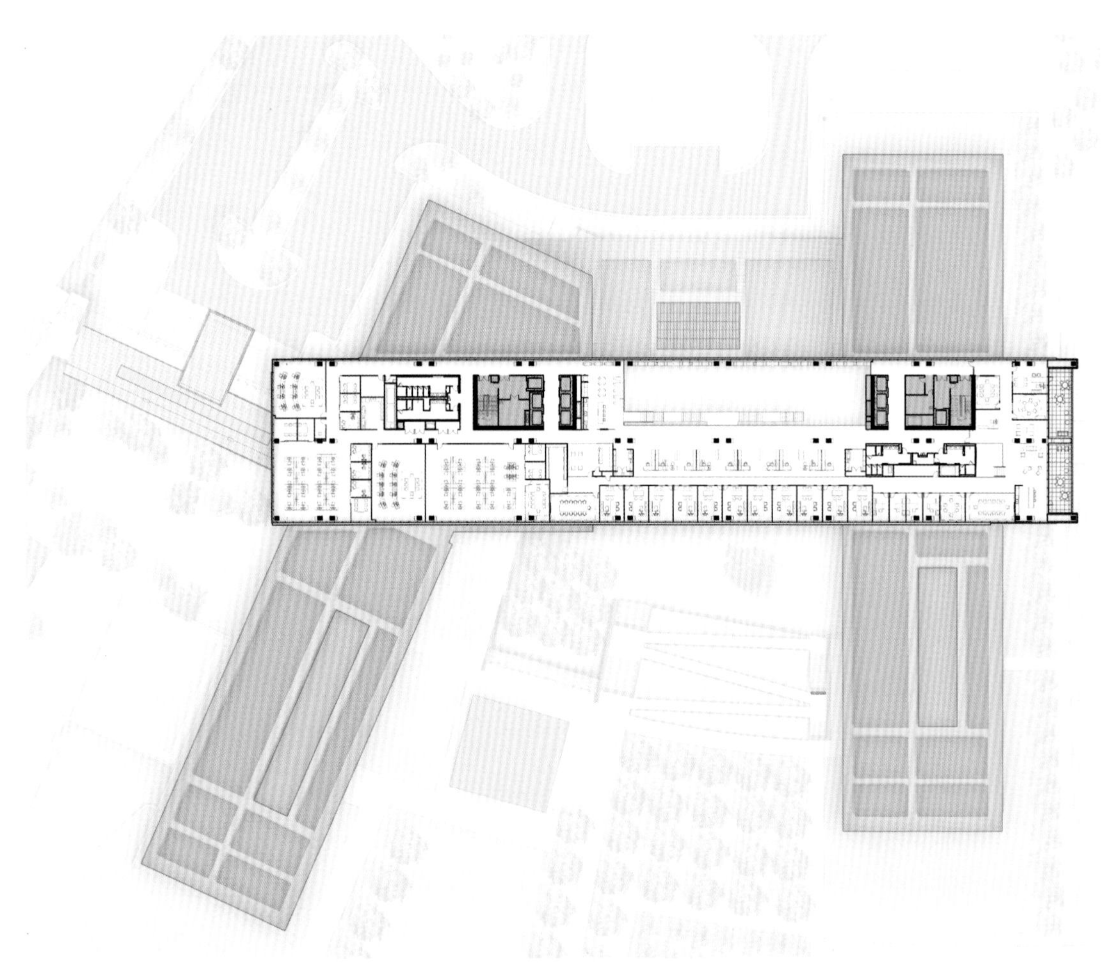

LEVEL 11 PLAN

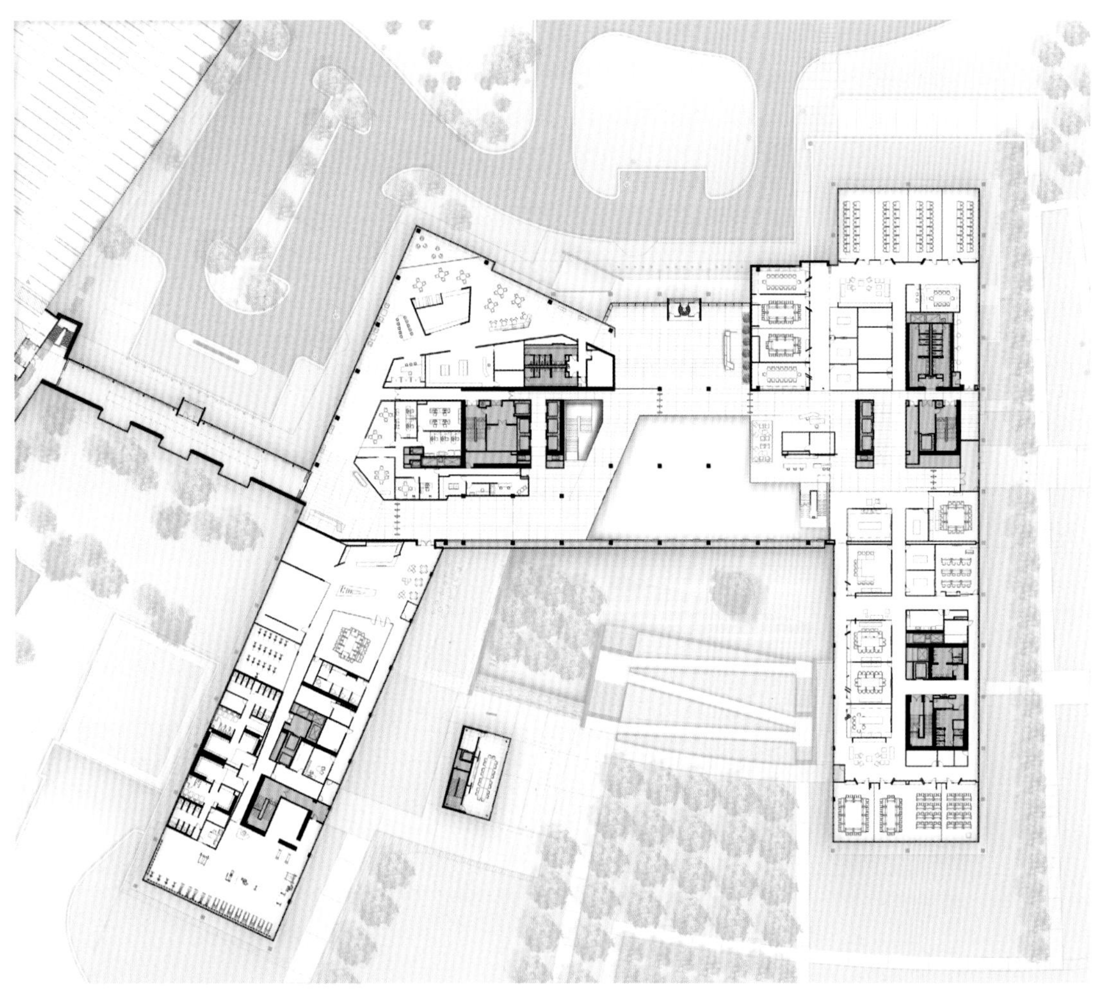

GROUND-FLOOR PLAN

ZURICH

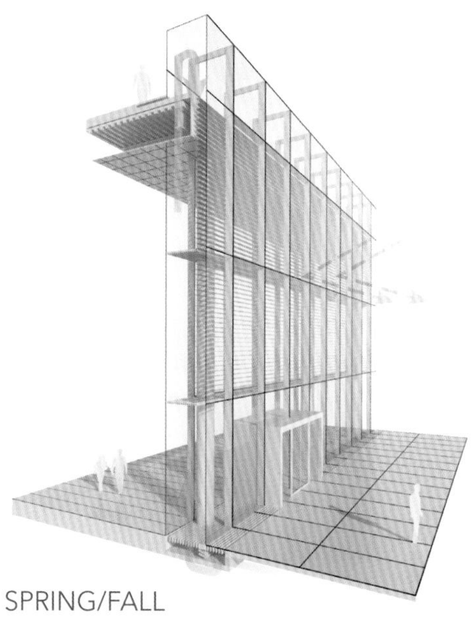

SPRING/FALL

VENTS OPEN/CLOSE AS NECESSARY TO MODULATE INTERIOR TEMPERATURE

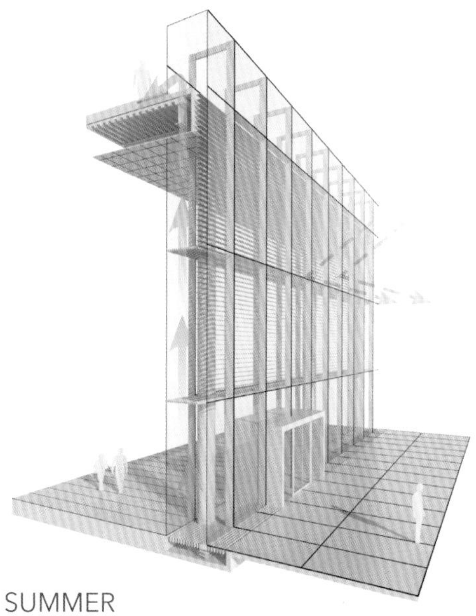

SUMMER

CAVITY IS HEATED BY SOLAR GAIN, CREATING A STACK EFFECT THAT DRAWS IN COOL AIR BELOW AND EXHAUSTS WARM AIR ABOVE

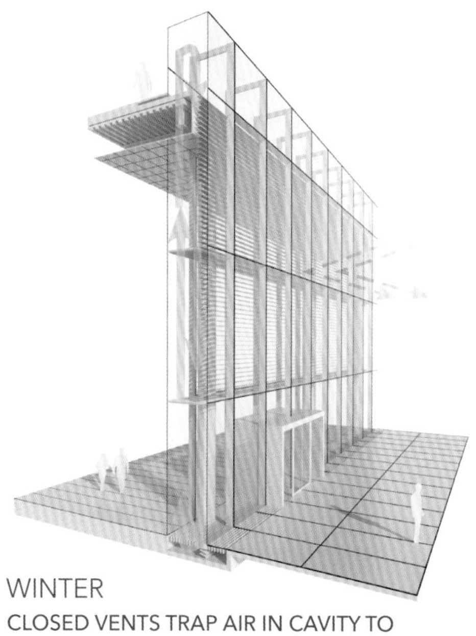

WINTER

CLOSED VENTS TRAP AIR IN CAVITY TO CREATE A WARM THERMAL BARRIER

阿布扎比AL HILAL银行办公大楼

AL HILAL BANK OFFICE TOWER

Abu Dhabi, United Arab Emirates

This 24-story office tower serves as the flagship commercial development for Al Hilal Bank, a progressive Islamic bank in Abu Dhabi. Located on Al Maryah Island, the Emirate's designated new CBD, the building is designed to attract leading national and global companies with Class A office space of an international quality. The building features efficient, column-free floor plates, floor-to-ceiling glass, and the latest technology and amenities–features that are not widely available in the UAE market.

Challenged to define a distinctive image that would reflect the bank's progressive brand while also setting an international aesthetic, the design defines a bold, contemporary tower that shifts in massing as it rises. The podium contains retail space and a dramatic three-story transparent lobby to the north, with pedestrian arcades on the east and west. Three cubical masses sit atop the podium, stacked like shifted blocks. These forms derive their interest from recessed corners that are offset from each other and distinguish the tower from others on the island. In addition, the building's façade changes at the created voids to accentuate the shifted aesthetic. Orange accents highlight the voids while reinforcing the bank's branding both day and night.

Intended to convey a timeless, elegant image, the façade is composed of an aluminum-and-glass curtain wall system with glass and notched metal-spandrel elements and vertical glass fins that enhance the building's verticality while also providing subtle shading. Achieving an Estidama 1 Pearl sustainability rating, the tower offers maximum transparency, with floor-to-ceiling, high-performance glass providing spectacular views for occupants while significantly increasing interior daylight.

A landscaped park and reflecting pool along the building's western façade draw pedestrian traffic by creating an inviting, shaded urban space. Seating for tenants and visitors helps further complement the outdoor setting.

Stunning is the word that jury members used to describe this elegant composition.

–David Scott, Technical Jury Chair,
CTBUH Best Tall Building Awards 2015

استدامة
مصرف الهلال
al hilal bank

Maryah Tower

阿布扎比环球市场广场
ABU DHABI GLOBAL MARKET SQUARE

Abu Dhabi, United Arab Emirates

Set on a previously undeveloped island, the five-building, 290,000-square-meter Abu Dhabi Global Market Square complex is the anchor of the Emirate's new CBD. Comprising four office towers and the headquarters building for the Abu Dhabi Global Market, an international financial center, the complex creates a signature image for downtown while providing an efficient, premier office environment. The complex also promotes a public, walkable ground plane and emphasizes sustainability.

The distinctive new headquarters building is a landmark, four-level facility. Glass-enclosed with a roof the size of a football field, the building is raised 27 meters above a 49-meter-diameter water feature on massive granite piers. The four structural piers house the stairs, mechanical risers and service elements for the headquarters. The building projects an image of strength and solidity as it overlooks the water facing back toward the city's existing downtown.

Four office towers frame the headquarters building: two at 31 stories and the other two at 37 stories. The first full office floor of each building starts 34 meters above the ground level, providing a highly transparent, open lobby and elevating the views on all tenant floors. At plaza level, the buildings sparingly touch the site, and indoor and outdoor spaces seamlessly blend to create a large, landscaped, pedestrian-friendly environment that connects the complex. Beneath the plaza, a two-story retail podium weaves through the development, providing upscale shopping along the waterfront.

The project emphasizes a sustainable design approach throughout. One of the main initiatives involves the environmentally responsive enclosure system that uses a mechanically ventilated cavity and a double-skin façade system over large portions of the office buildings. These elements help mitigate the 40 °F (22.22 °C) interior-exterior temperature differential. The double-skin cavities run the entire height of the four office towers. Within the cavities, active solar shades track and adjust for the sun angle in order to provide optimal shading to the building's interior. Active lighting controls also help balance natural and artificial light. Because of these and other measures, the complex became the first integrated development in Abu Dhabi to be certified LEED-CS Gold.

This is a very powerful composition where sustainable features have been embedded from the design stage as part of the core functions within the complex.

—Karen Weigert, Juror, CTBUH Best Tall Building Awards 2013

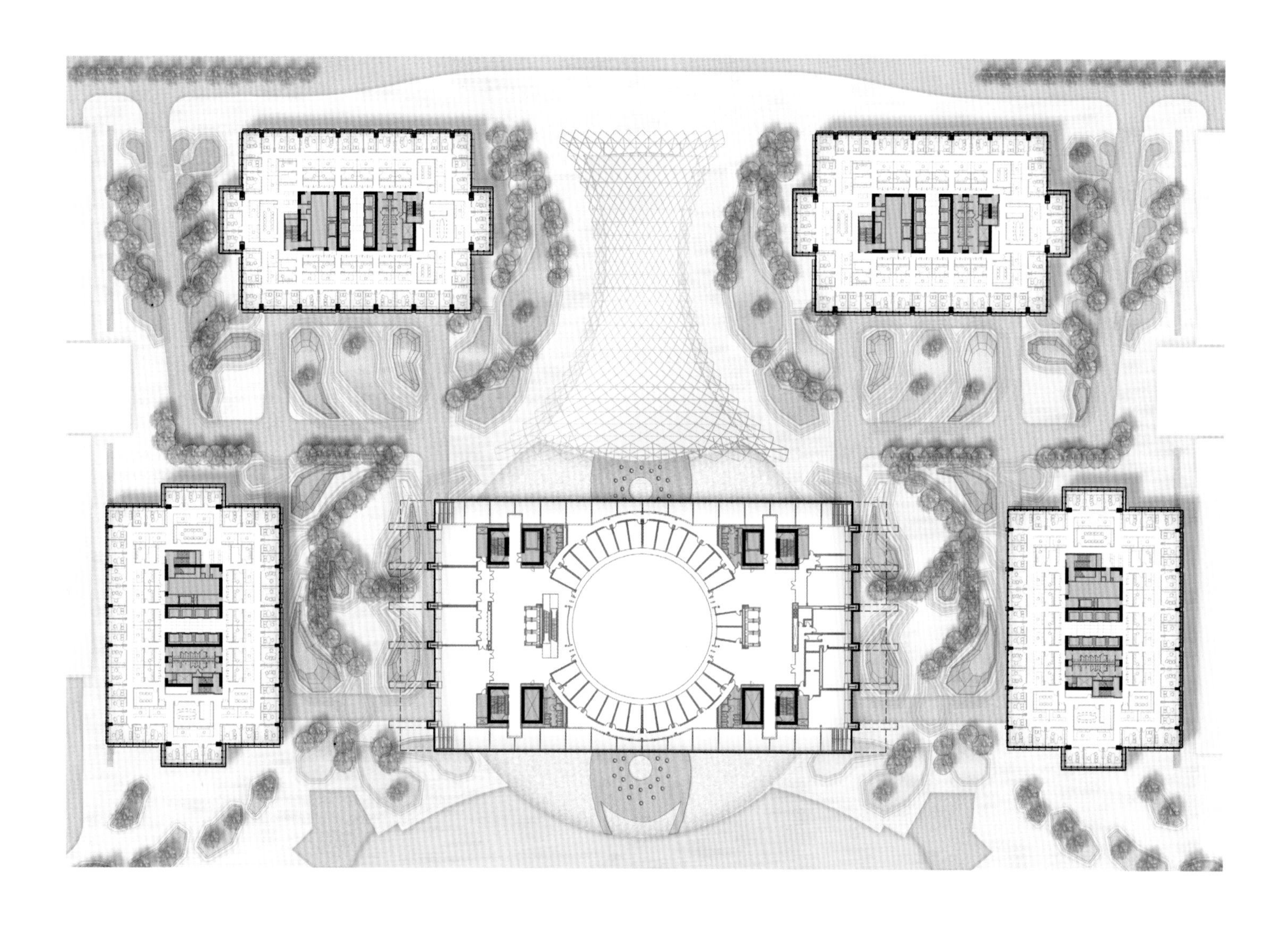

南京市麒麟科技创新园
NANJING QILIN TECHNOLOGY CAMPUS HEADQUARTERS

Nanjing, China

The Nanjing Qilin Technology Campus Headquarters is an office building that blends civic dignity and authority with unity and openness, an image befitting its combination of governmental and commercial functions. The structure consists of six levels above grade and two below-grade parking levels.

The basic L-shaped plan of the building is composed of four components. A rhythmic series of commercial office "fingers" connected by transparent atria defines the building's north edge. To the south, an L-shaped wing of government offices forms the urban corner and provides a dominant building image. The south wing embraces a lower volume containing a conference center, exhibition halls, and a food and beverage center. These three major programmatic and massing elements are then linked by a pristine glass entrance hall, together conveying a formal order and transparency that identify the structure as a government building. The entrance hall is located at the nexus of a linear axis defined by an exterior reflecting pool to the north and internal pre-function spaces to the south. These components form an elegant yet monumental structure that establishes an edge presence along East Guang-Hua Road while also opening toward the technology park to the west.

The project is a low-energy and environmentally friendly structure. Green roofs, water pools, rooftop photovoltaics, and façade shading help the complex integrate the natural environment. The linear form of the building is well suited for office space, offering flexibility for a variety of tenants. Optimized daylight illumination, healthy conditioning, the incorporation of garden atria, and spectacular views from all workstations ensure a vibrant commercial office environment.

At the heart of the project lies the desire for a building that blends civic dignity and authority with unity and openness.

–Hui Zhang, Vice President, R&F Properties

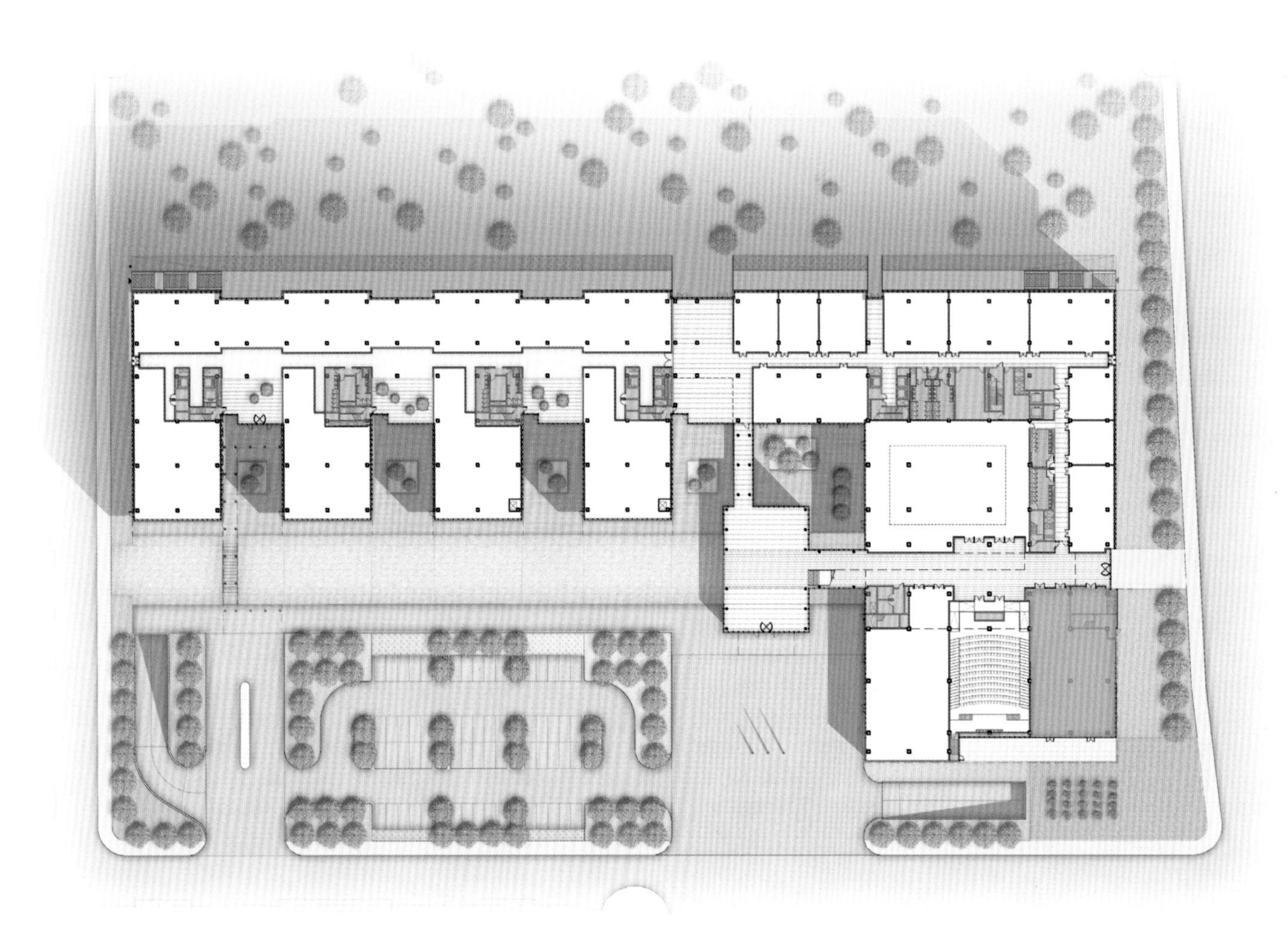

东吴证券总部大楼

SOOCHOW SECURITIES HEADQUARTERS

Suzhou, China

Within Suzhou, China, the office and stock exchange building is sited along the western edge of Jinji Lake in Suzhou Industrial Park, a major commercial and mixed-use district outside the city center. Prominently anchoring a major boulevard that leads to central Suzhou, the building is conceived as a modern gateway to the historic city center. The triangular massing design responds contextually to the view corridors of the city and lake, the solar orientation of the site, and the major diagonal vehicular artery. The triangular form is also seen symbolically in China as a balanced and stable form, an image well suited to a stock exchange headquarters.

The signature feature of the design is a soaring internal atrium that rises the full height of the building and is enclosed with a dramatic, engineered triangular skylight. The interconnectivity of the atrium creates a commanding presence, allowing access to natural light and views from all locations within. The building enclosure is designed to minimize the overall energy consumption of the building. The triangular building form creates a self-shading massing that minimizes the east and west exposure, allowing for easier control of solar gain along the south façade. A high-performance shingled tower façade is employed to provide passive shading during the warmest summer months.

The drama of this building's monumental atrium brings a grand sense of scale that belies its modest height.

—Georges Binder, Managing Director, Buildings & Data

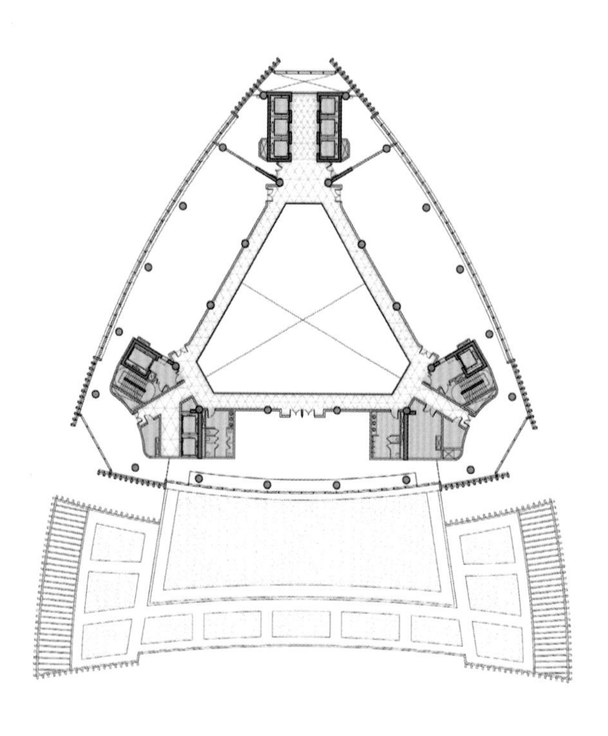

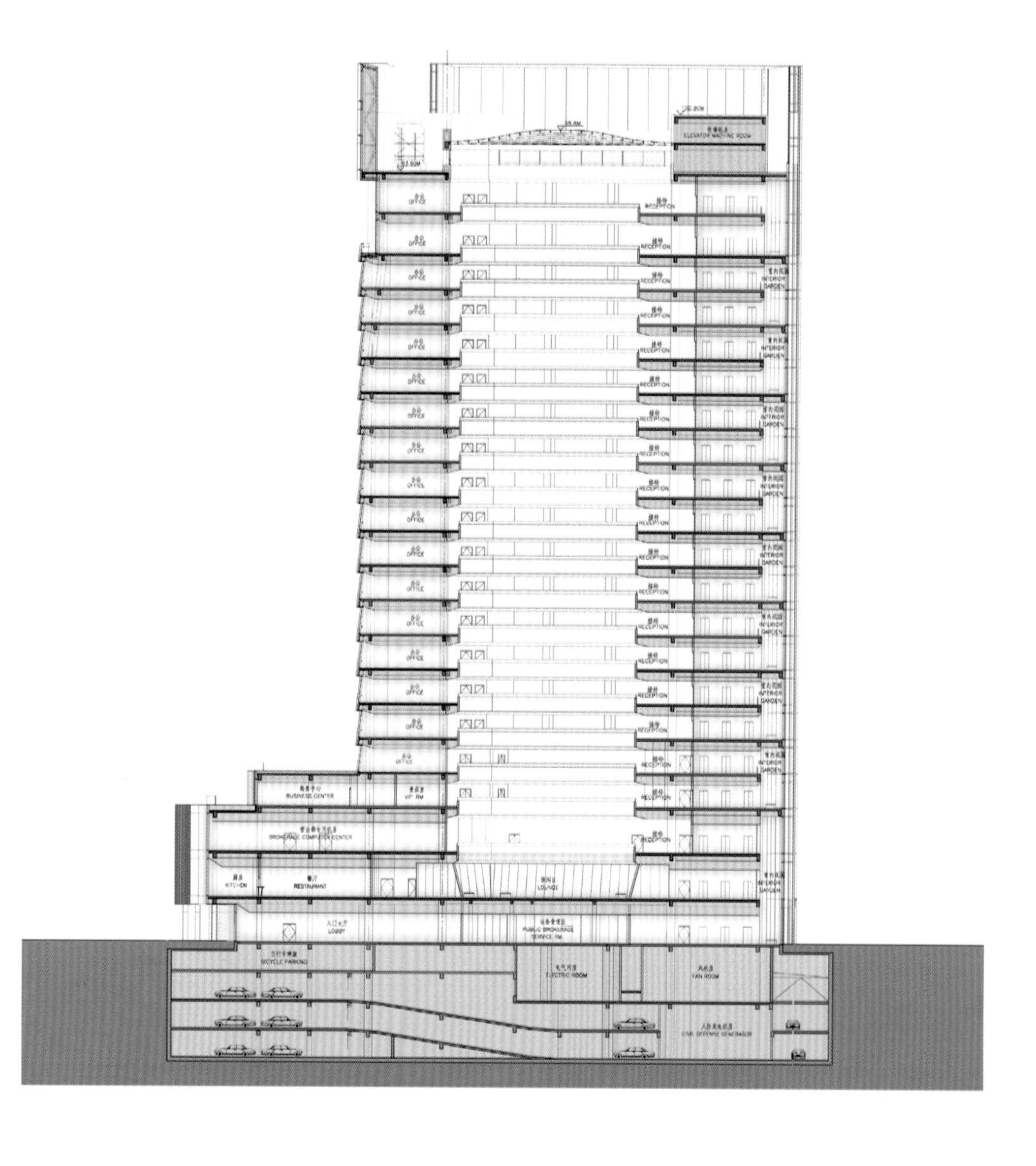
OFFICE
RECEPTION
INTERIOR GARDEN
BUSINESS CENTER
VIP RM
BROKERAGE COMPUTER CENTER
KITCHEN
RESTAURANT
LOUNGE
LOBBY
PUBLIC BROKERAGE SERVICE RM
BICYCLE PARKING
FAN ROOM

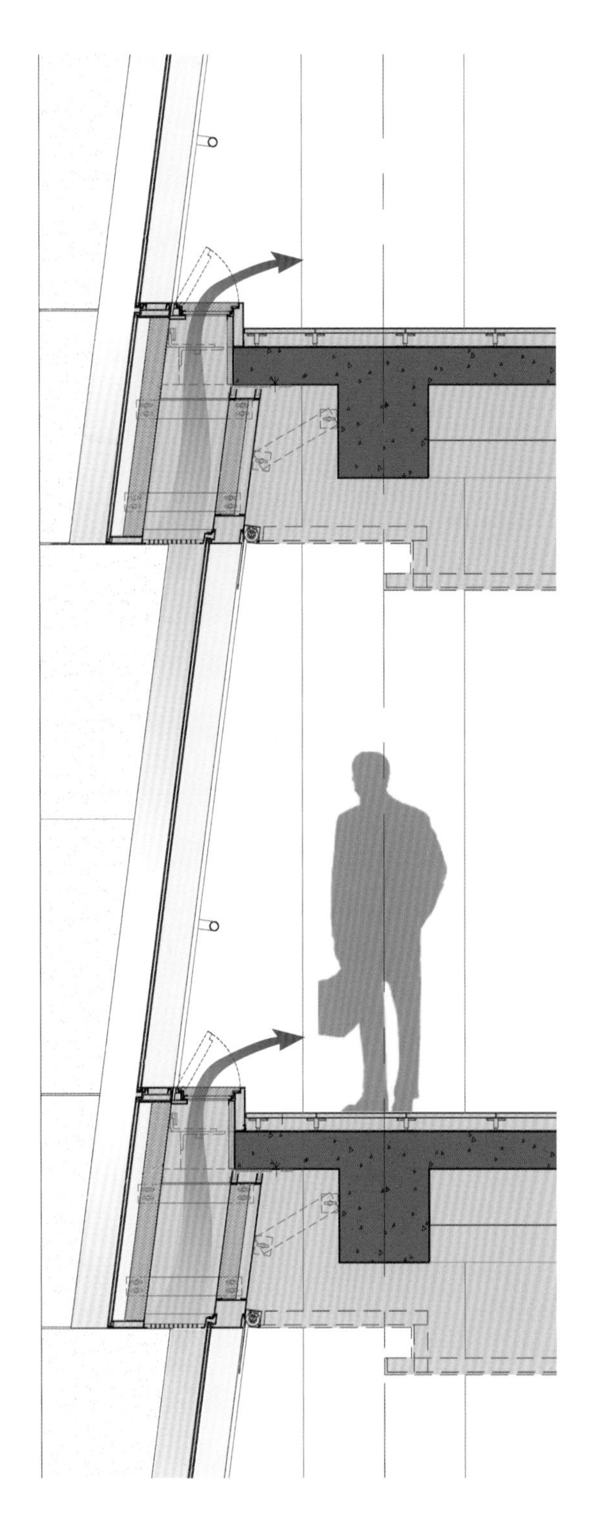

富力盈凯广场
R&F YINGKAI SQUARE

Guangzhou, China

Located in Guangzhou's new city center of Zujiang, R&F Yingkai Square emerges as part of a larger master plan of mixed-use towers that collectively signify the stature of Guangzhou as a major metropolitan city. The simple yet distinctive form of the tower traces inspiration from the abundant local bamboo plants, rising 296 meters and defined by the building's asymmetrically carved corners as well as the veining of vertical strips on the façade that provide a sense of visual movement. The strips compress and stretch as they rise, starting more dense at the base to enhance the sense of gravity. The Park Hyatt Guangzhou hotel occupies the building's uppermost floors, with office space below and subway connections below grade.

While the tower internalizes its functions into a singular expression, the design is greatly born of its context. The square tower massing respects the geometric rigidity of the street grid, helping to form urban rooms in conjunction with the neighboring parcel. The pinching language created by carving out the corners highlights the unique views available at various heights through and over neighboring structures, while the diagonal extensions of the site relate to the adjacent central green and nearby Pearl River Delta.

The building is at once connected to the urban street life around it while balanced with an intimacy required for a luxury hotel experience. The hotel drop-off and arrival sequence is choreographed to emphasize a sense of calm, with a warm, neutral palette punctuated by sculptures that aid in orientation. An infinity-edge pool on level 60 runs the length of one side of the building, overlooking the panoramic views of the Guangzhou skyline. A signature outdoor roof garden on level 70 offers similar sweeping views while providing an inviting space to dine and relax.

The tower establishes a defining silhouette in Guangzhou's rapidly evolving skyline.

–Hui Zhang, Vice President, R&F Properties

ASCOTT
雅诗阁服务公寓
YE 越秀集团
ASCOTT

ASCOTT

群光广场
CHICONY SQUARE

Chengdu, China

This 37-story mixed-use building is designed to provide an identifiable architectural expression through its unique façade treatment and bold massing along the fast developing Chengdu skyline. The project occupies a full city block adjacent to a vibrant, pedestrian-only public square in the heart of Chengdu's shopping and dining district. As a mixed-use building, the project is anchored by a 12-story department store, with a 25-story Grand Hyatt hotel above. Together, these components provide 111,500 square meters of retail space, 370 guestrooms, and supporting hotel amenities.

Visually active retail entries are directly accessed from the pedestrian-only avenues to the north and west, while the more tranquil hospitality entries are organized to the south and east for intuitive vehicular access. Siting the hotel asymmetrically to the south emphasizes the overall verticality of the tower, while ensuring that all rooms have unobstructed views and minimizing the shadows cast on the active urban plaza to the north. A series of stepped gardens atop the podium are integral to the architectural union between tower and podium. These urban terraces are accessible from the hotel's public areas and help provide an architectural character completely unique in downtown Chengdu.

The primary material palette of opaque white glass and high-performance vision glazing establishes a dramatic contrast within the urban context. The façade design is organized by a series of architectural "planes and reveals" that provide scale to the project, while the shifting windows and joint patterns create a sense of visual movement across the taut exterior surfaces. Contemporary retail "awnings" are formed by cantilevered glass volumes that activate the building base. These cantilevered volumes provide a continuous rhythm of canopies for pedestrians while organizing extensive signage and advertising needs.

The white glass of this tower creates a distinct image in Chengdu. The building will have a lasting impact for many years to come.

—Dr. Kun-Tai Hsu, Chairman, Chicony Electronics Co.

正熙国际酒店

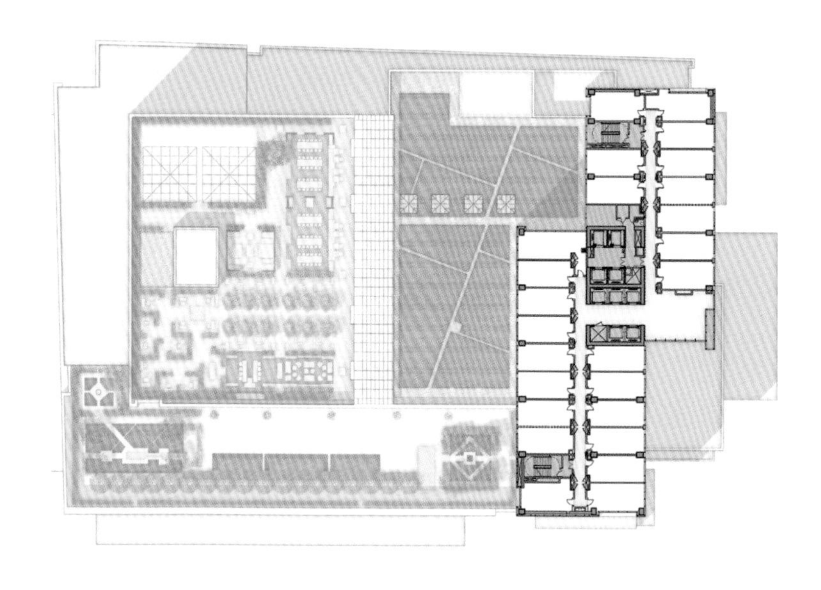

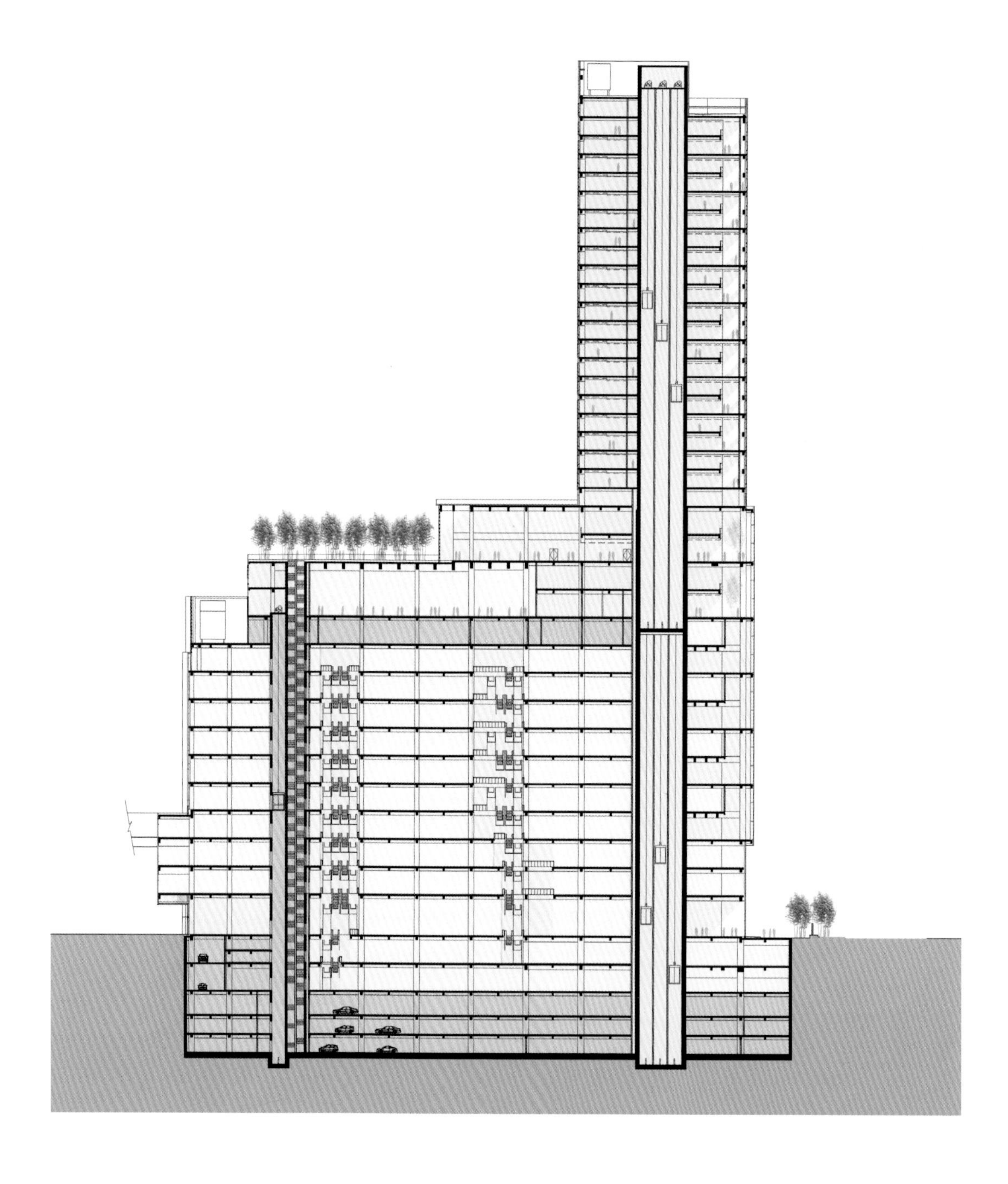

大连君悦酒店
GRAND HYATT DALIAN

Dalian, China

Located in China's northeastern coastal city of Dalian, this 370-key hotel and 84-unit serviced apartment tower sits unobstructed overlooking the Yellow Sea to the south and China's largest public plaza, Xinghai Square, to the north. Reflecting its unique site, the building's design responds to both its physical and environmental context. The tower's triangular plan ensures that all rooms receive southern light as well as views of the sea and nearby mountain ranges. Additionally, the triangular form helps to mitigate the impact of the area's uniquely high winds, channeling them to the building's northern corners and minimizing the structural impact on the slender tower.

Programmatically, the hotel floors are stacked below the serviced apartment levels, enabling the core to telescope and creating the architectural "portal" along the north façade. Internal circulation is exposed on this face to provide all vestibules with ambient light and views of the park and skyline, while assuring a consistent lighting profile at night. The top two levels of the tower house the signature restaurant, offering unobstructed views in all directions.

The tower sits atop a four-level podium that houses large banquet, meeting, dining, fitness, and spa facilities. The main dining facility connects the building physically to its beachfront location while other functions, including the entry plaza and lobby, are elevated to capture commanding views of the bay and adjacent park. The fitness center and spa open out to a series of south-facing terraces, which provide both afternoon sun and a visceral connection to the sea.

The tower's façade is composed of floor-to-ceiling, high-performance glazing to maximize the views, and horizontal metallic sunshades are sized for solar protection and privacy. In contrast, the building's four-level podium is clad in a mixture of warm limestone, stainless steel, and high-transparency glass that relates to the scale and materiality of the neighboring residential villas.

This project started with a simple goal: to give every room a view of the sea.

–Rong Liu, Vice President, China Resources Land Limited

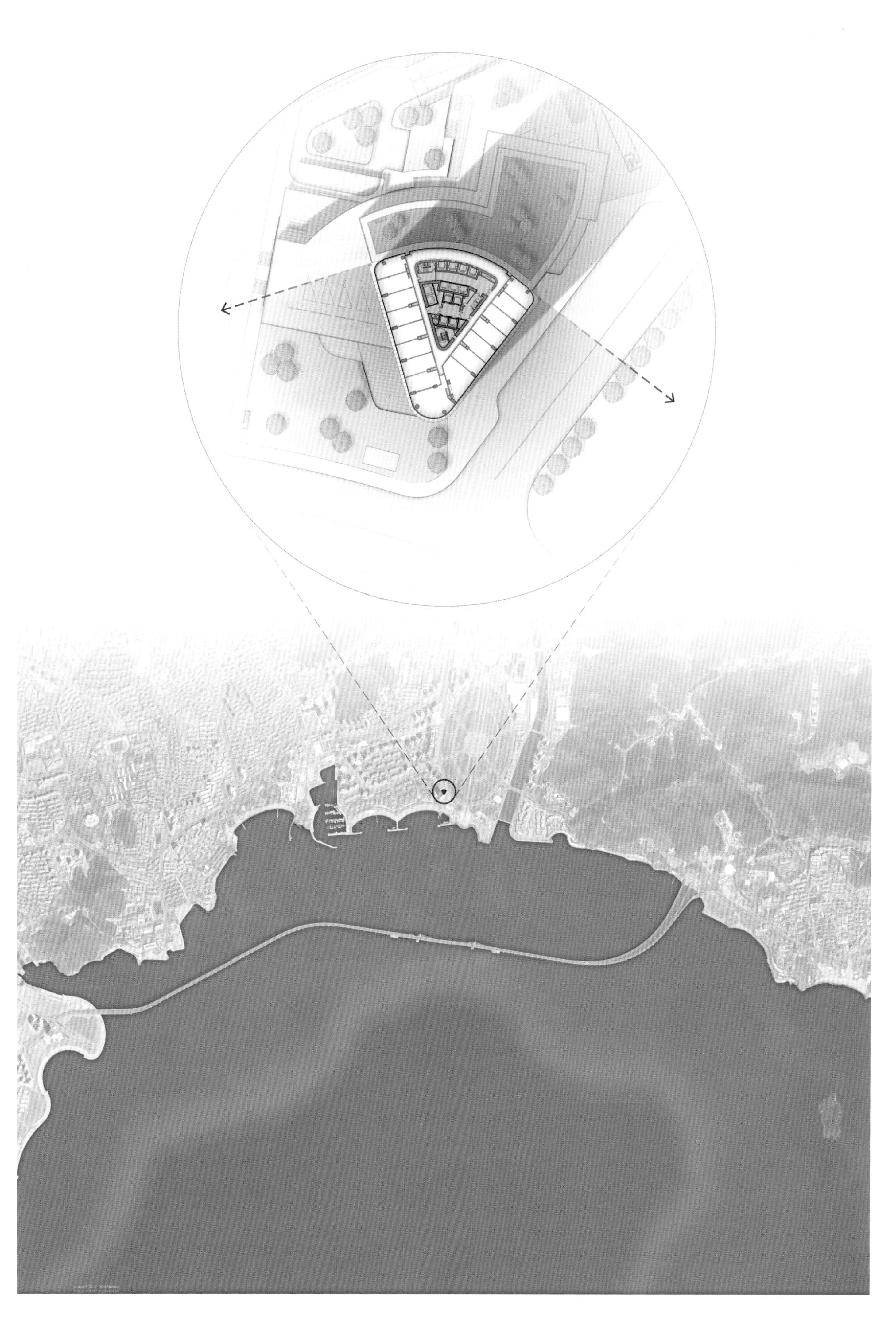

GRAND
HYATT
君悦酒店
GRAND
HYATT

GRAND HYATT

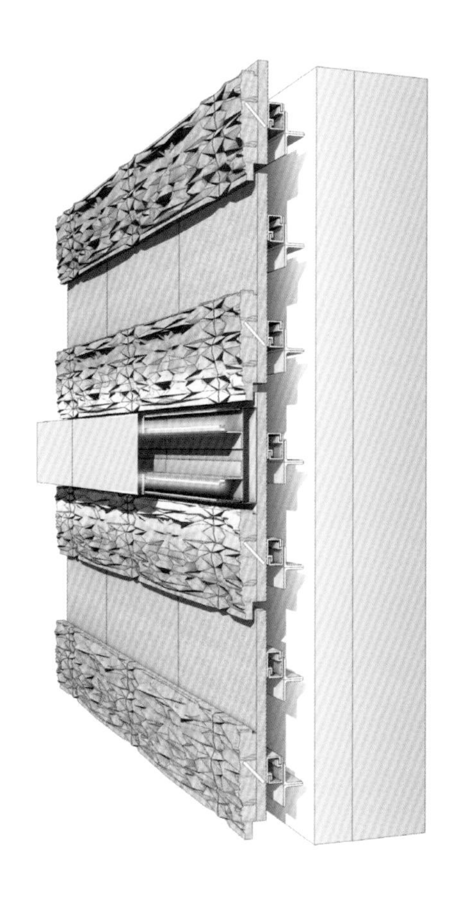

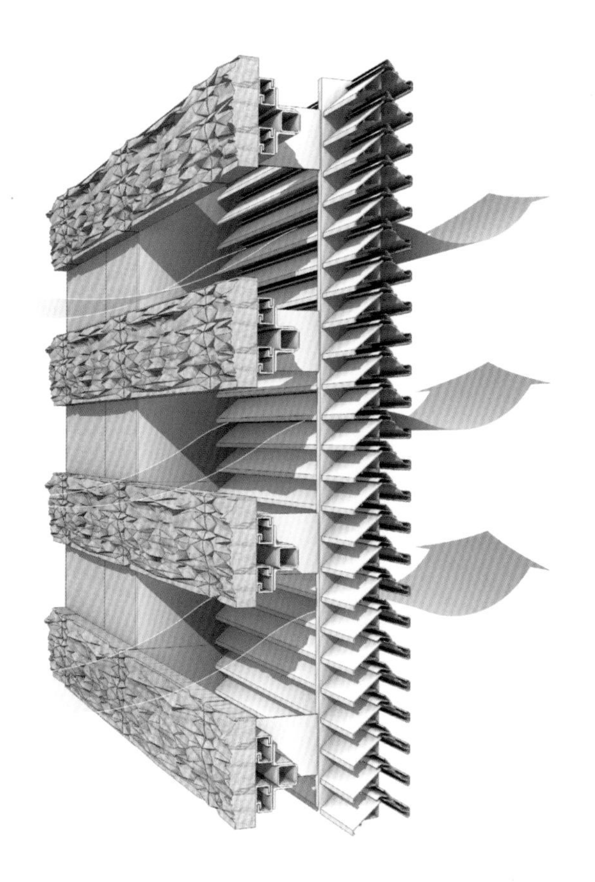

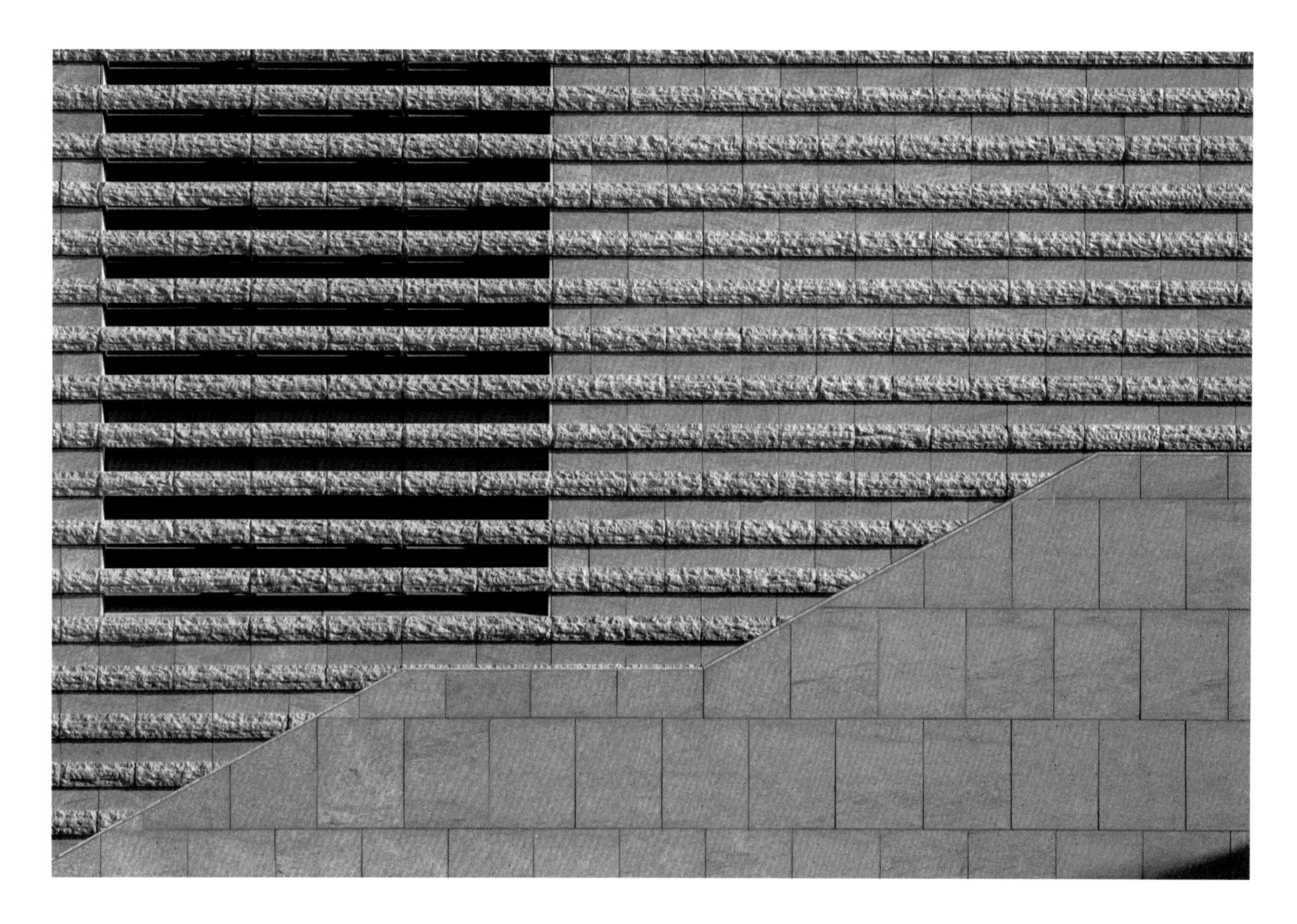

深圳湾木棉花酒店
HOTEL KAPOK SHENZHEN BAY

Shenzhen, China

Located adjacent to the Shenzhen Bay Sports Center athletic complex, the new Hotel Kapok is a 19-story boutique hotel with 242 rooms. The design attempts to provide guests with a unique spatial quality that eliminates the cloistered and disorienting nature of typical hotel corridors. The resulting pinwheel plan allows for corridors that end in vision glass with a punctuation of light and views. Additionally, four three-story atriums rhythmically ascend up and around the hotel core to drive light deeper into the floor plans while strategically organizing amenity areas vertically throughout the building. Each atrium is located to celebrate views that are best by direction and height: south and low, to the adjacent pedestrian plaza; east and high, to distant Hong Kong Bay; and north and west in the middle, for views of the surrounding athletic venues.

The hotel's exterior expression is unique, clad entirely in high-performance glazing, overlaid with a perforated metal scrim that provides shading and references the similar diamond-shaped pattern on the enclosure of the adjacent sports center. The four atriums reveal themselves on the outside as large voids in the scrim. A rooftop garden and restaurant offer expansive views of the city, bay, and Hong Kong beyond.

This building has a very dynamic master plan, and the use of the diagonal as a unifying element both horizontally and vertically was very effective.

—Juror, AIA Chicago Distinguished Building Awards 2015

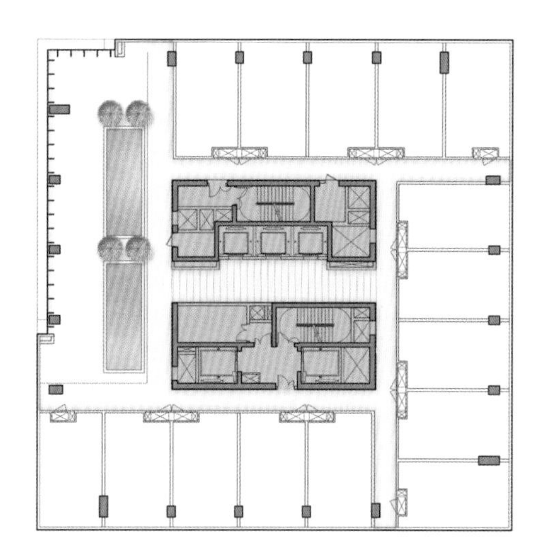

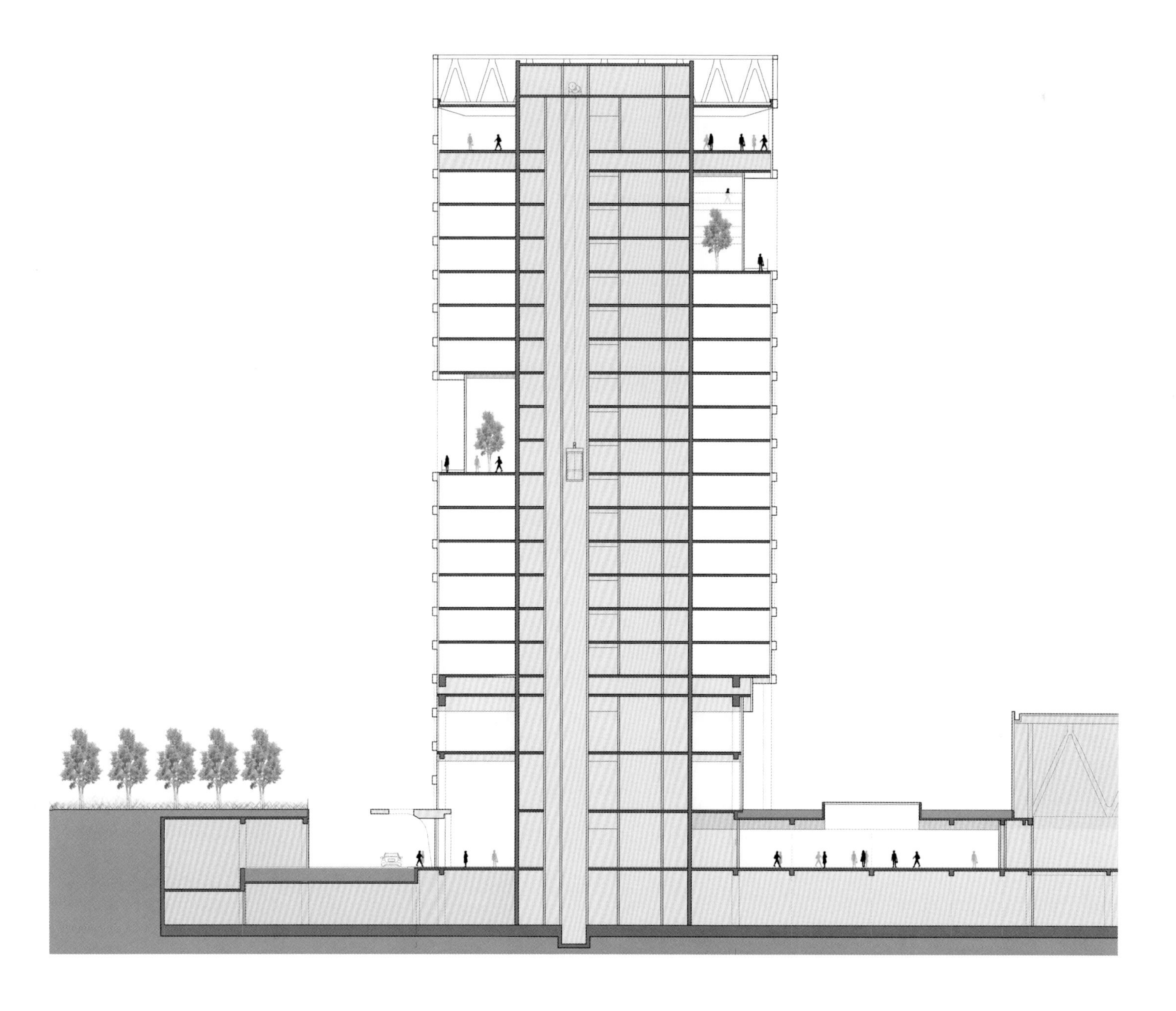

杭州钱江新城万豪酒店
HANGZHOU MARRIOTT HOTEL QIANJIANG

Hangzhou, China

Located along the western bank of Hangzhou's Qianjiang River, this complex features a 348-key Marriott hotel, as well as a 51-unit serviced apartment tower. A four-level podium connecting the angular twin towers provides house banqueting, meeting, dining, fitness, and spa facilities.

The epicenter for the project, the main hotel lobby, is a multistory space under the south tower. The all-day dining restaurant, lobby lounge, specialty restaurant, and hotel bar are located along the eastern perimeter of the building–on levels one and two–to ensure that views of the river and gardens are maximized, while the grand ballroom, junior ballroom, and meeting facilities are stacked vertically along the western façade.

The unique tower floor plan shape references the fractured geometries of the neighboring gardens. Their linear and stepped profiles ensure that most rooms and suites have views of the gardens and river to the east.

High-performance glass, silver-metallic accents, and dark, textured stone are employed in the exterior design. Inspired by the wave patterns of the river to the east, the geometry of the main glass façades creates a "ripple" effect across the longitudinal façades of the towers. This expression promotes verticality while obscuring the cellular nature of the hotel rooms contained behind. Additionally, the rippled edges of the curtain wall create shading devices that enhance solar protection and privacy while maintaining floor-to-ceiling glass to maximize views. The dark, textured granite surfaces of the building provide a visual grounding in contrast to the light and reflective glass surfaces of the towers.

The complex's design celebrates the unique characteristics of the sloping site through thoughtful placement and shaping of the programmatic elements. A centered, terraced garden forms a focal point and links the distinctive buildings together in a cohesive campus design.

For an urban hotel, the Hangzhou Marriott has a remarkable resort-like sense of arrival.

–Georges Binder, Managing Director, Buildings & Data

MARRIOTT
MARRIOTT

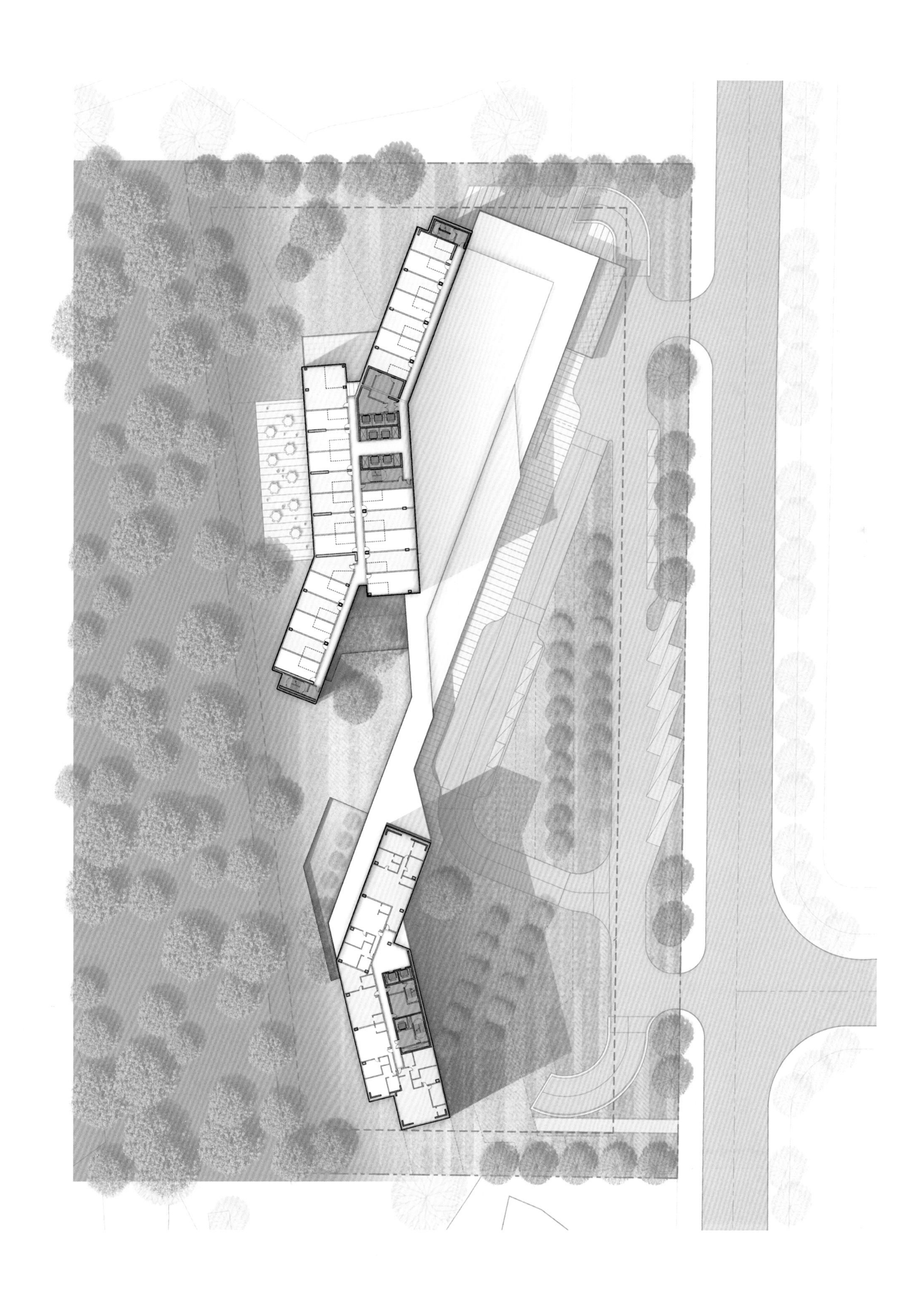

MARRIOTT

芝加哥总督大厦
VICEROY CHICAGO

Chicago, Illinois, USA

Located at the intersection of Cedar and State streets in the heart of Chicago's Gold Coast neighborhood, the historic Cedar Hotel takes on a new identity as the first Viceroy property in Chicago. The four-story brick-and-terra-cotta façade of the original Cedar Hotel is preserved and given fresh life, fronting a new 18-story hotel tower. Visitors enter an open three-story lobby space with an adjacent restaurant and lounge. Additional amenities include a ballroom and conference suite on level four, with access to a large outdoor roof terrace overlooking State Street. The rooftop lounge, with an outdoor pool and landscaped terrace, offers views of Lake Michigan and the Chicago skyline.

A unique identity for the Viceroy emerges through the contrasting façades of the new tower and the historic building. The brick and terra cotta of the 1920s-era Cedar Hotel exemplifies the vintage character of a building façade as a two-dimensional arrangement of building elements and ornamentation. By contrast, the glass curtain wall of the new hotel tower is a taut, three-dimensional envelope. The distinctive "folding" geometry is inspired by the argyle-like pattern found in the historic brick façade below. The design solution creates "harmony through contrast" in which both the old and new are rich examples of their specific eras.

With 180 guestrooms and suites, Viceroy Chicago is a reborn property at a high-profile intersection in one of Chicago's most vibrant and rapidly changing neighborhoods. In combination with the hotel's other amenities, the existing plaza will be reactivated with outdoor dining that further complements the one-of-a kind guest experience.

The combination of old and new is a bold solution and distinctive addition to an active neighborhood in the city.

–David B. Nelson, Head of Real Estate, Convexity Properties

VELVET TACO
STATE
RUSH

VICEROY
VICEROY
VICEROY

芝加哥希尔顿伦敦之家酒店
LONDONHOUSE CHICAGO

Chicago, Illinois, USA

Located in downtown Chicago at the prominent intersection of Michigan Avenue and Wacker Drive, LondonHouse Chicago is an adaptive reuse project that brings together old and new. A 1923 office building, originally designed by Alfred Alschuler for the London Guarantee & Accident Company, is combined with a new slender infill tower on an adjacent site to define an integrated 452-key hotel. The renovation encompasses 305,000 square feet of the existing 21-story building, with an expansion of 85,000 square feet next door.

Completing the street wall by filling in a previous parking lot, the new tower is designed to respect the cornice lines of the existing property while introducing a series of subtle saw-toothed angles in the new façade that respond to the signature views west down Wacker Drive and overlooking the Chicago River.

The hotel's main entrance, located in the new glass tower, features a gateway arrival leading to the historic elevator lobby and a grand second-floor check-in lobby and bar. The top of the hotel is the main feature. Previously unoccupied, underutilized space has been converted into a tri-level rooftop terrace and bar, including a special-events space within the refurbished cupola. Designed in keeping with the city's landmarks codes and approvals, the rooftop space sensitively inserts glass rails stepped back from the original façade to allow the historic building details to shine while providing guests exceptional views of downtown and along the river.

There's nothing quite like watching a beautiful old building snap back to life after years of neglect.

–Blair Kamin, Architecture Critic, *Chicago Tribune*
"Michigan Avenue Classic Comes Back to Life with a 21st Century Twist," May 26, 2016

WYNDHAM
LONDONHOUSE
CHICAGO WATER TAXI

LONDONHOUSE

箭牌大厦
THE WRIGLEY BUILDING

Chicago, Illinois, USA

The Wrigley Building, designed by Graham, Anderson, Probst & White, is one of downtown Chicago's most recognized architectural icons, dating to the 1920s. Following the sale of the property in 2011, new ownership sought to renovate the building, retaining its name and restoring its historic integrity while positioning it to serve a new mix of 21st-century tenants. The most historically sensitive work focused on the building's exterior, lobbies, and plaza, while floors above were completely renovated to serve new office users.

One of the more significant efforts was the removal of the screen wall between the two towers at the ground level, which also involved structural reframing and terra cotta restoration. The 1920s plan for the building had anticipated an upper-level street that would run between the towers. Although this street was never built, the removal of the glazed screen and the 1950s connecting walkways accomplished the 1920s vision of creating an open passage and plaza.

Work on the plaza itself was extensive. The entire 1950s-era plaza was demolished down to structural steel. The area was rebuilt using new pavers in a consistent color and materials palette, and the heritage bronze storefront was wrapped along both sides. The redeveloped plaza defines a distinguished, large open space that presents an inviting outdoor amenity for passersby and caters to prospective retailers and restaurants.

Inside the towers, major public areas were also renovated. In particular, within the building lobbies, low ceilings, remnants from historically lacking 1980s renovations, were removed and replaced with sympathetic interpretations of the original designs, restoring the original volumes and utilizing 1920s marble and mahogany. Corridors were largely retained, including marble walls and floors, as well as original doors. In addition, nearly all of the building's windows—more than 2,000—were replaced, and MEP and life safety systems were either replaced or modernized to serve incoming tenants for many years to come.

It is one of those projects that was a labor of love—a Chicago icon meticulously brought back to life for a new generation of tenants.

—Ari Glass, President, Zeller Realty Corporation

帕特里克 · G. 与雪莉 · W. 瑞恩音乐艺术中心
PATRICK G. AND SHIRLEY W. RYAN CENTER FOR THE MUSICAL ARTS

Evanston, Illinois, USA

Northwestern University's new music and communication building is located on the southeastern edge of the Evanston campus, sitting on a prime site fronting Lake Michigan. The design of this world-class facility is a response to the university's goal to provide a dramatic architectural statement that signifies the public importance of the School of Music and optimizes views of the lake. Designed to wrap and connect with an existing 1970s music building, the new Ryan Center enables the School of Music to consolidate its programs for the first time ever, completing a more than 40-year vision.

Guided by the university's master plan, the 152,000-square-foot building forms a new quadrangle with the other fine and performing arts facilities to provide a physical identity for the collaborative arts community. The dynamic, Z-shaped plan of the building mass defines the eastern edge of a new arts green then jogs to the western edge of the lake. One enters the building through a large atrium, which serves as a main circulation and gathering space and offers striking lake views. The atrium is the link that connects the performance spaces with the practice and teaching rooms, and it provides a visual connectivity between floors that helps activate the space.

Building materials relate to other campus facilities, with the three main performance venues–recital hall, opera/black box, and choral–all clad in limestone. The classrooms, offices and practice rooms are enclosed with a unique double-skin façade designed to provide each practice room with absolute acoustical separation. Certified LEED-NC Gold, the building offers a variety of innovative sustainable features, including the double-skin façade, a greywater system, and other initiatives.

The 400-seat recital hall is the building's signature venue. A 40-foot-high double-skin glass wall provides the performers with a dramatic backdrop of Lake Michigan and the Chicago skyline. To address acoustical challenges of the monumental glass wall, the inner glass slopes inward, eliminating reverberation. Horizontal woven-wood panels on the other three sides further absorb reflections while lending a feeling of warmth. The entire room works in harmony to provide a dramatic visual and acoustical setting for world-class performances of all kinds.

It is, on the whole, an impressive design: sensitive to its surroundings, but strongly sculptural.

–Blair Kamin, Architecture Critic, *Chicago Tribune*
"Northwestern's Music Building in Harmony with Lakefront"
October 13, 2015

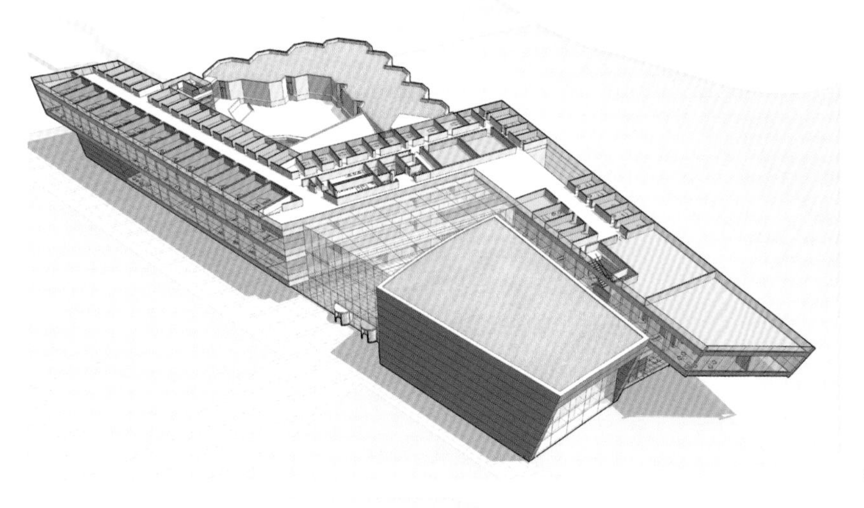

ADMINISTRATIVE-LEVEL PLAN

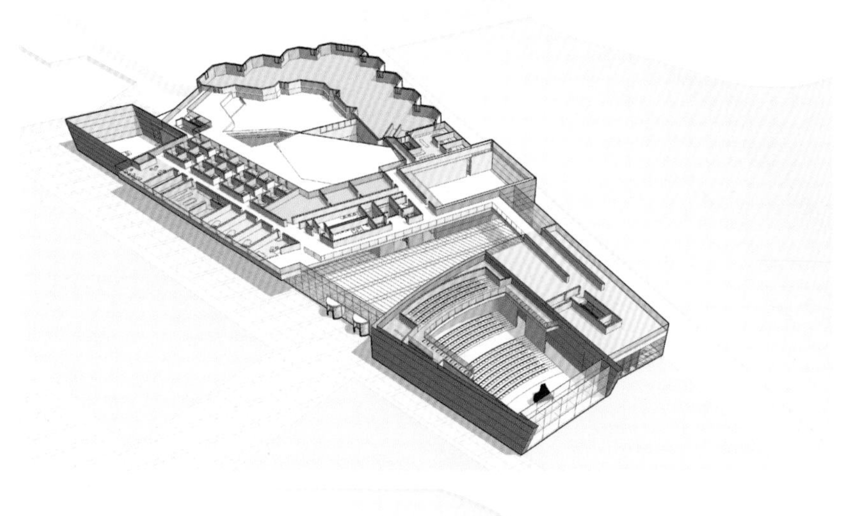

PRACTICE-LEVEL PLAN

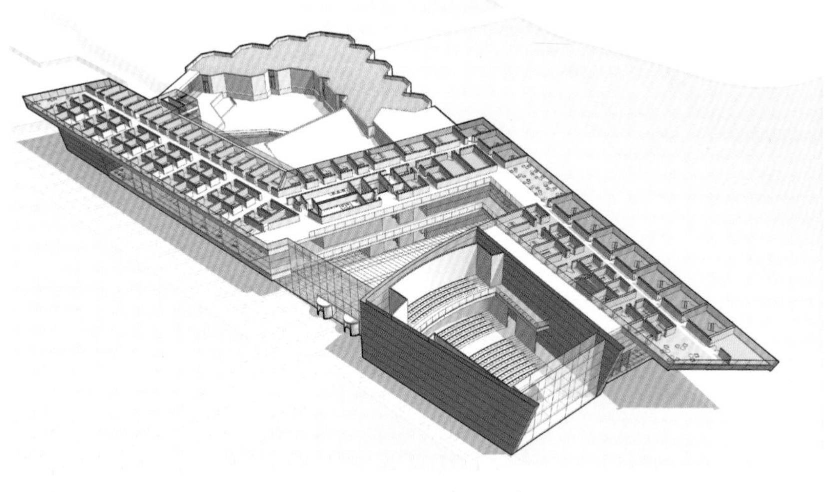

TEACHING-LEVEL PLAN

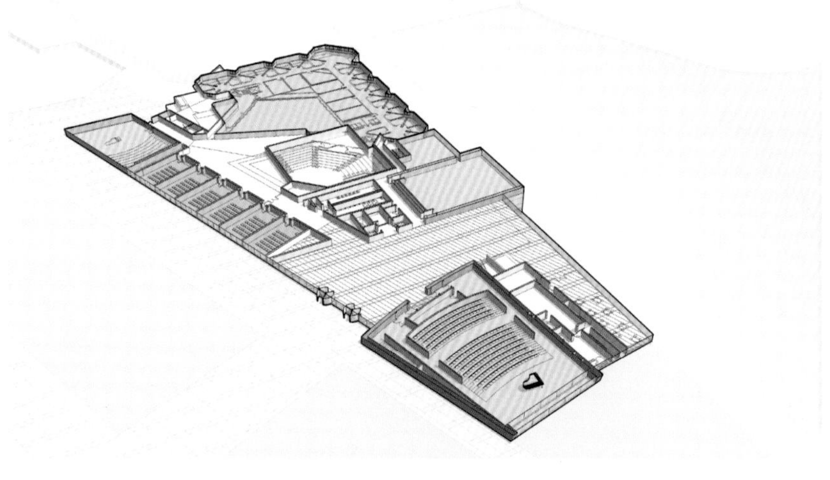

PERFORMANCE-LEVEL PLAN

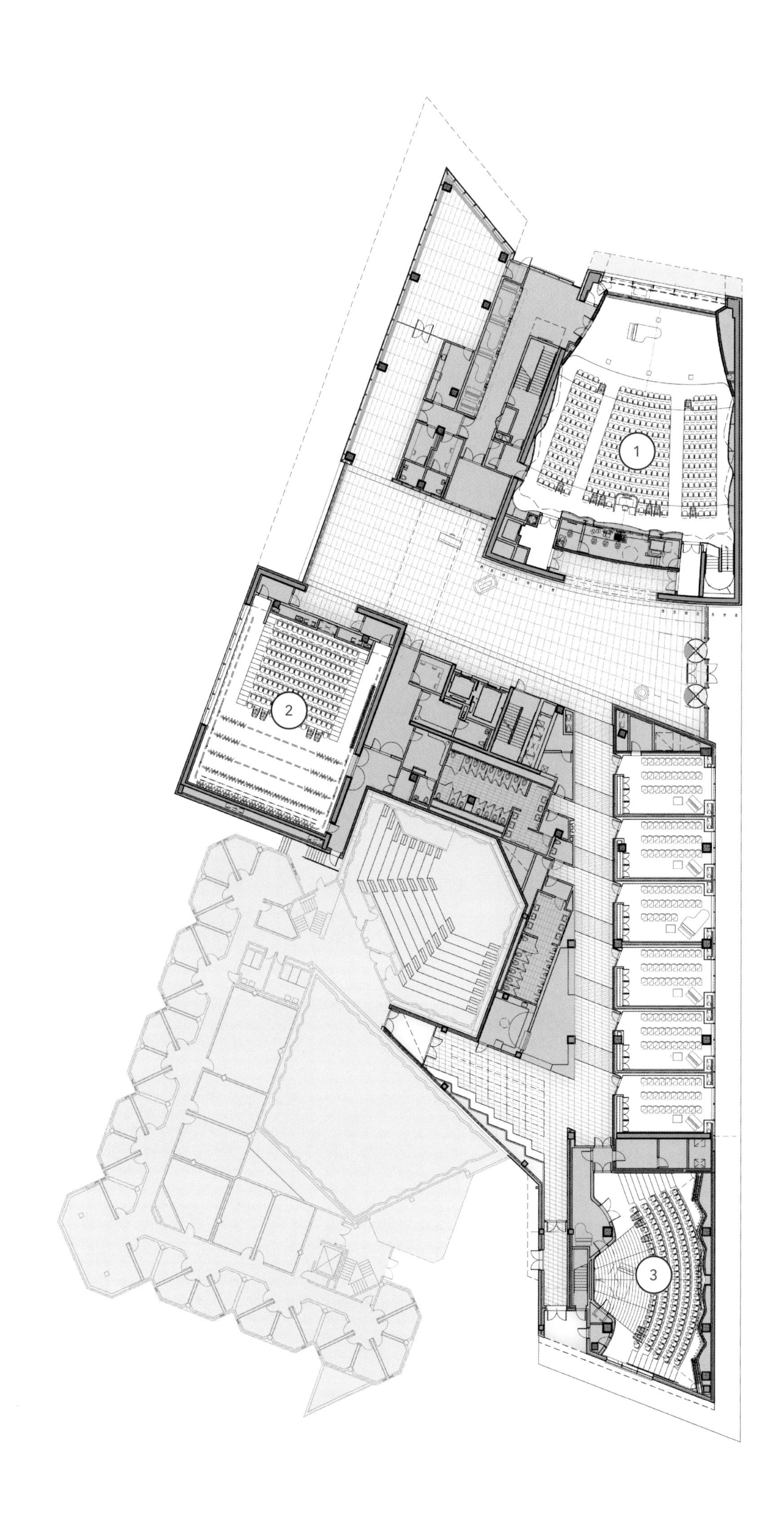

1 RECITAL
2 BLACK BOX
3 CHORAL

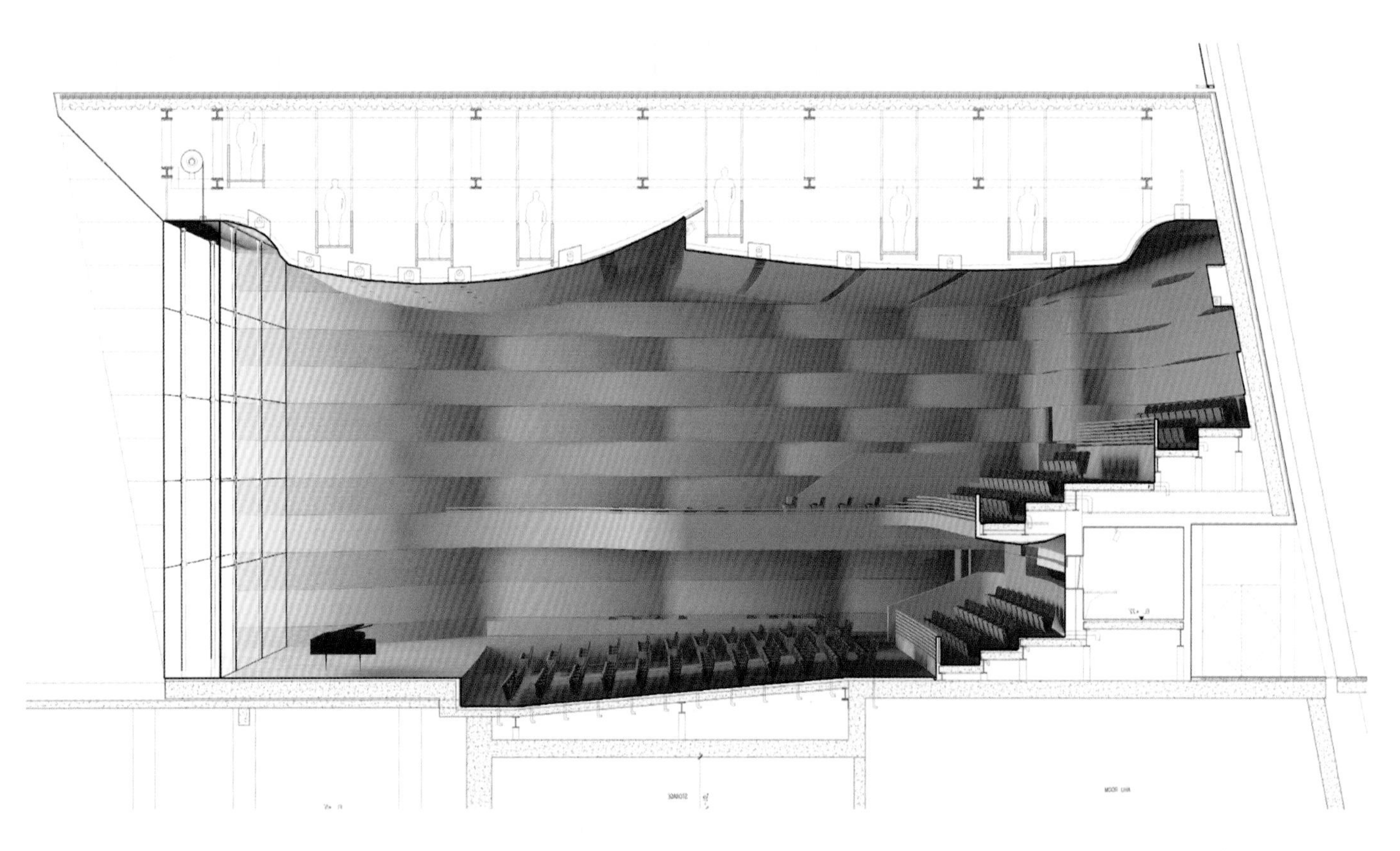

在建及未来项目
ON THE BOARDS

伟基河畔北110号
110 NORTH WACKER

Chicago, Illinois, USA

Offering one of the best office building locations in Chicago in terms of accessibility and visibility, this site would have been developed many years ago if not for its trapezoidal shape. Given that the east and west property lines are not parallel, fitting a typical Class A office floor plate, including the center core and desired lease spans, is a challenge.
In addition, for any site along the river, the city requires a 30-foot-wide publicly accessible riverwalk, further complicating the site.

The design addresses the challenges with an unusual stepped center core, which allows for a 45-foot lease span on each side. To avoid an angled perimeter, the west wall has a series of 30-foot-wide, 5-foot setbacks, accommodating an orthogonal, 5-foot planning module throughout the building.

The state-of-the-art Class A office tower totals 54 stories and 1,650,000 square feet. Interior space planning is enhanced by the 5-foot setbacks along the river, providing the equivalent of 14 corner offices. The setbacks create a distinctive form that accentuates the building's verticality and avoids the pure box-like appearance of typical office towers.

To satisfy the city's requirements, the design provides a 45-foot-wide riverwalk that is covered, but effectively open to the sky, 55 feet above. Three large, tree-like structural elements spaced 90 feet apart transfer out the tower columns along the river and further open the site. The result is a landscaped, covered walkway that connects two very important pedestrian paths and maintains an effectively unobstructed river view. In combination with other space, the site is 50 percent open and publicly accessible at grade.

On the Wacker Drive side, a 45-foot-high lobby is enclosed by a cable-supported glass wall, and sculpted, folded-limestone cladding covers the elevator cores. The transparency of the low-iron glass wall virtually eliminates the distinction between interior and exterior space, with the lobby and streetscape becoming one.

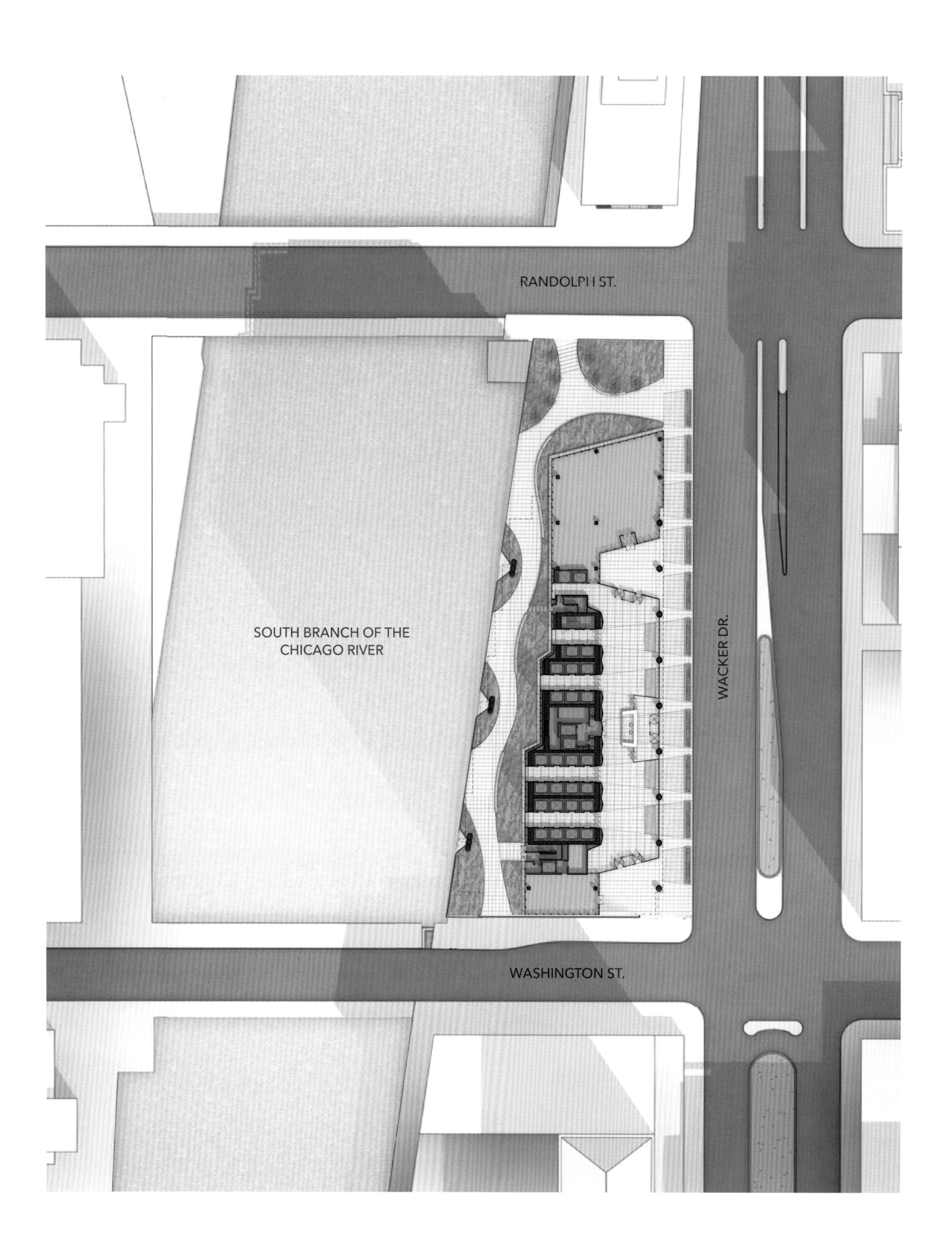
RANDOLPH ST.
SOUTH BRANCH OF THE
CHICAGO RIVER
WACKER DR.
WASHINGTON ST.

跨海湾帕克大厦
PARK TOWER AT TRANSBAY

San Francisco, California, USA

Located at the corner of Howard and Beale Streets in the Transbay district of San Francisco, this commercial office tower sits across the street from the new Transbay Transit Center and just three blocks from San Francisco Bay. The 45-story tower is designed with three massings that offer a variety of floor plates. Within these distinct massings, the design carves out a series of large outdoor terraces that provide substantial outdoor amenity space for tenants–mini "parks in the sky."

The tower's façade is composed of floor-to-ceiling glass and is articulated with vertical glass fins that relate to the location of the tower terraces, which occur at a three-story module. The podium façade also utilizes a series of single-floor fins to break down the scale at the pedestrian level. The fritted pattern is repeated through to the base of the building to help continue the tower articulation and materiality throughout the three tower massings.

At the base of the tower is a covered 3,600-square-foot outdoor plaza with a 35-foot-high ceiling. The area is designated as privately owned public outdoor space (POPOS) that abuts both the tower's main lobby and adjoining retail space. Along with a series of outdoor seating areas, a major focal point of this outdoor space will be a large, site-specific artist installation, which will be integrated into the main wall of the tower's core.

With design features such as raised floors, natural ventilation, and high-efficiency building systems, the project is designed for LEED Gold certification.

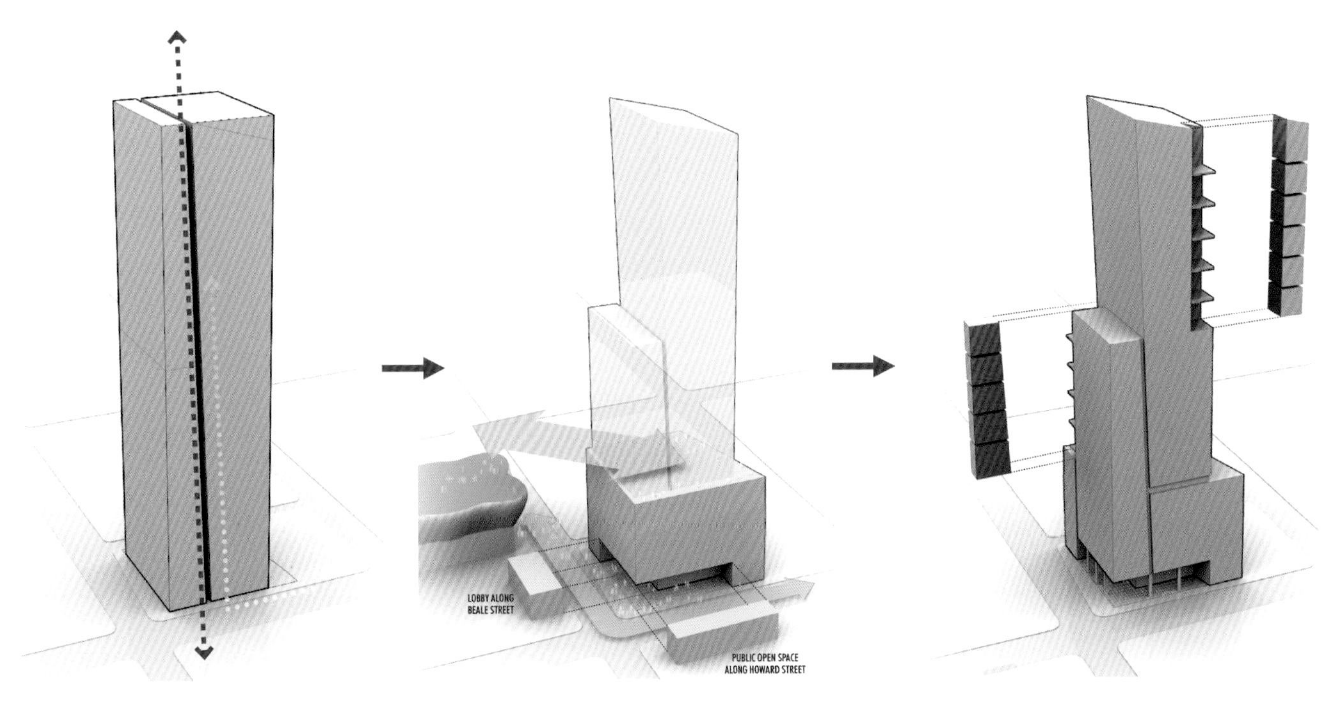

BEGIN WITH SIMPLE EXTRUDED FORM AND BREAK DOWN MASS WITH DIAGONAL SEPARATION

CREATE LOBBY AND OUTDOOR PUBLIC SPACE AT BASE WITH A PODIUM MASS THAT MAINTAINS CONNECTION TO CONTEXT

OUTDOOR TERRACES ARE CARVED AWAY FROM MASSINGS

迪拜综合大厦
DUBAI MIXED-USE TOWER

Dubai, United Arab Emirates

Located at the Za'abeel roundabout in Dubai, this mixed-use tower combines a hotel, serviced apartments, office space, and luxury condominiums. The site is on a figurative fault line between the old and new sections of Dubai. The tower's east-west axis aligns Za'abeel Palace, symbolizing old Dubai, with the Union House, representing the new. This alignment provides the primary rationale for the building massing and placement.

Rising to a total height of 305 meters, the tower is defined by an oval footprint that unites two slender, independent volumes with a large void on the upper levels. A series of atria with bridges and balconies fill the void in the lower and upper portion of the tower, creating an "urban window" that allows for views both of and from the building. In addition, this "window" frames dramatic views of the sunrise and sunset through the building along the east-west axis.

The atrium is envisioned to become the central circulation space for the hotel and office floors. Panoramic elevators and transparent glass bridges that connect the two sides of the building make the atrium a vibrant, active space with unparalleled views of Dubai. At night, the atrium becomes a glowing lantern with exterior lighting designed to provide an abstract representation of the lunar calendar.

On the exterior of the tower, vertical double fins provide shading while adding a rich texture. This articulated façade distinguishes the transparent, light-filled void of the atrium from the more solid detailing of the tower's exterior shell.

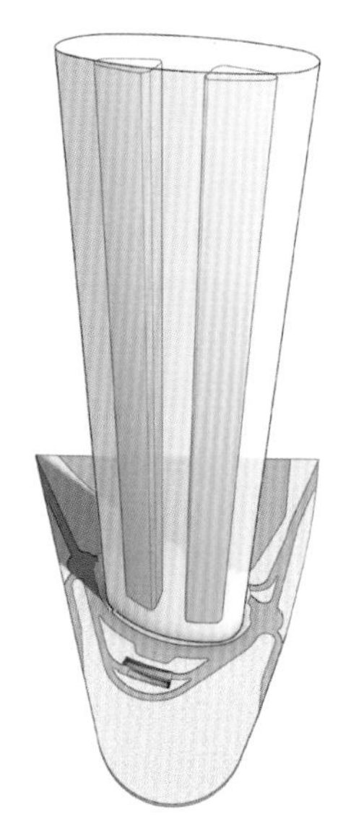

TWO TOWER CORES ARE UNIFIED TO CREATE ONE CONTINUOUS FORM; THE BUILDING IS THEN ORIENTED IN RESPONSE TO THE SITE CONSTRAINTS AND SURROUNDING CONTEXT

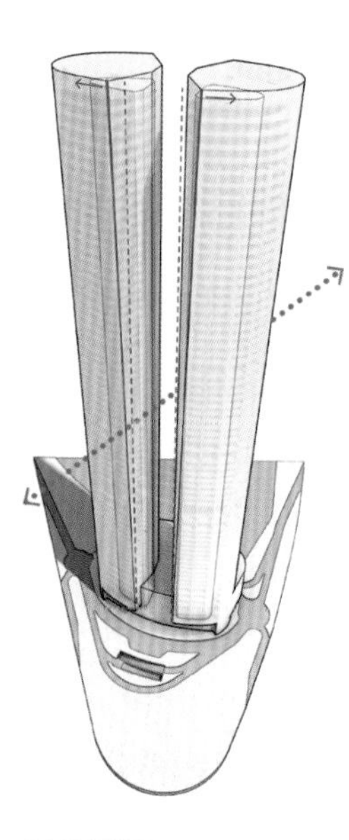

A LARGE VOID IS PROJECTED THROUGH THE MASS OF THE BUILDING TO EMPHASIZE EAST-WEST VIEW CORRIDOR

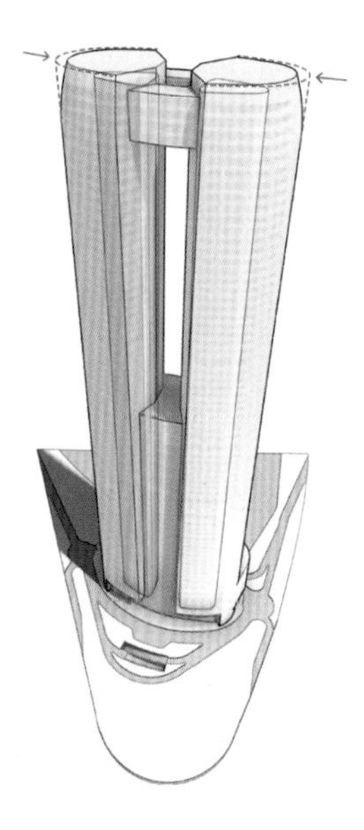

A SERIES OF ATRIA WITH BRIDGES AND BALCONIES FILL THE VOID IN THE LOWER PORTION OF THE BUILDING; THE TOP OF THE BUILDING ALSO GENTLY CURVES INWARD TO BRING THE TWO FORMS OF THE TOWER BACK TOGETHER

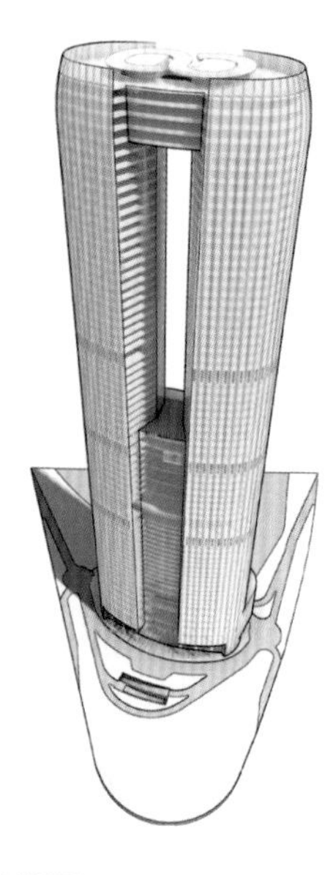

VERTICAL DOUBLE FINS PROVIDE SHADING WHILE ADDING A RICH TEXTURE TO THE FAÇADE

利雅得希尔顿酒店
HILTON RIYADH HOTEL & RESIDENCES

Riyadh, Saudi Arabia

Designed in collaboration with Omrania & Associates, Hilton Riyadh is located along the city's Eastern Ring Road, a major thoroughfare that links King Khalid International Airport with downtown. Two towers—a 20-story, 645-key hotel and a 14-story, 221-unit serviced apartment building—are connected by a large podium at grade. The podium provides a variety of food and beverage outlets, 6,500 square meters of meeting facilities, and a large 4,100-square-meter multipurpose hall, the largest in the city. Leisure and recreational facilities include multiple spas and health clubs, as well as a pool. The roof of the multipurpose hall features a mixture of garden areas and terraces that connect the building to the surrounding landscape, as well as relate to the area's overall master plan.

The project's two towers anchor the site on the east and west, with each planned and configured as three pinwheeling legs around a central elevator core. The towers are sited to optimize views and solar shading, as well as allow for the large multipurpose hall to span the area between the structures. This connection creates a relationship and orientation in the tower masses to unify the complex.

The hotel tower has multiple two-level atriums that rotate around the elevator core, creating internal light wells and providing movement in the overall massing. At the end of the guestroom corridors, windows offer daylight and additional views of the surrounding site, which aid in orientation. The hotel façade is expressed with undulating reveals along the glass faces, with solid, tapered end walls to frame the mass.

The apartment tower has linear slots along the intersecting three wings running the height of the tower to bring light into the core. Here, the glass façade has a weave of undulating forms that, again, bring movement to the mass and are framed by solid, tapering end walls.

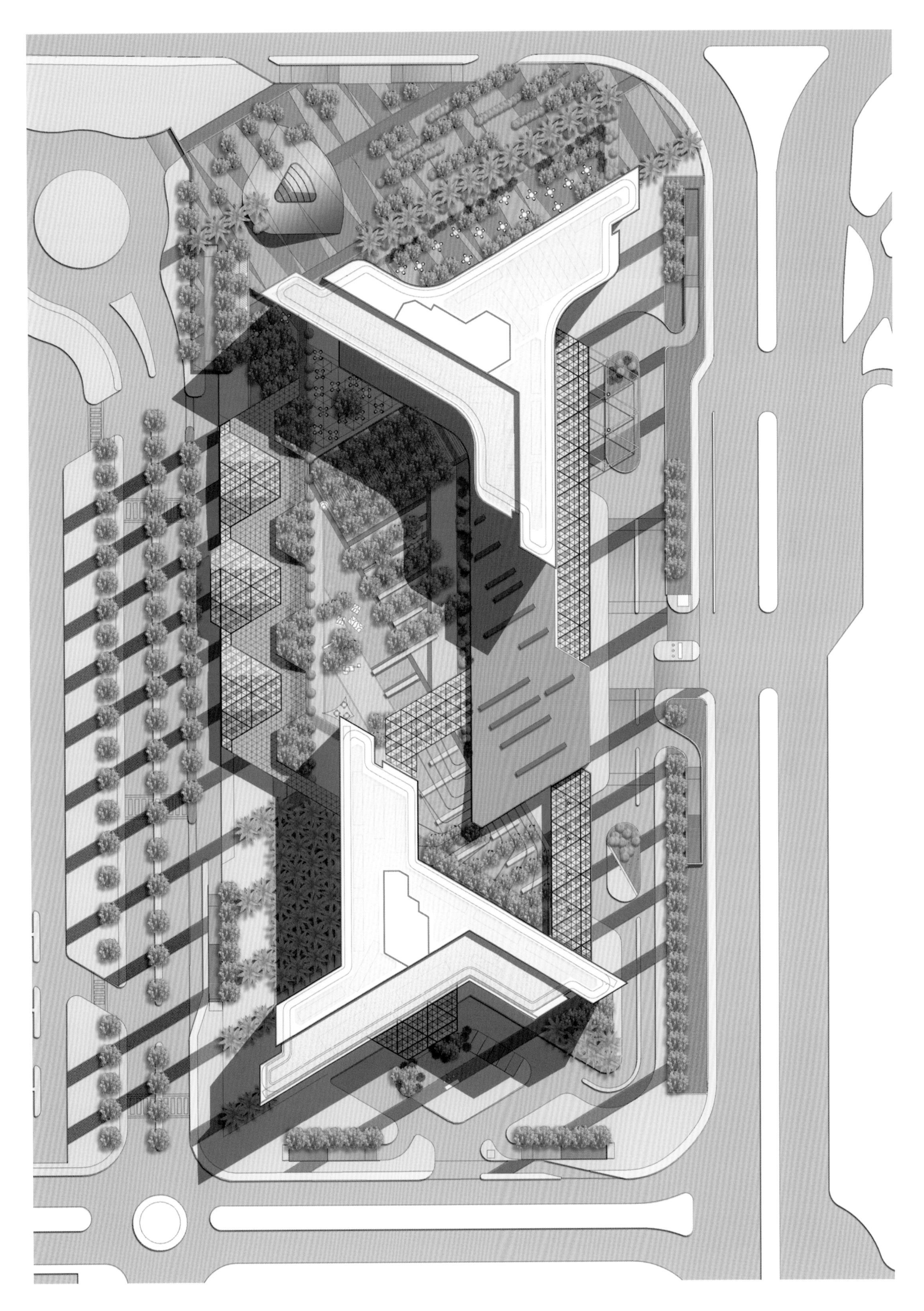

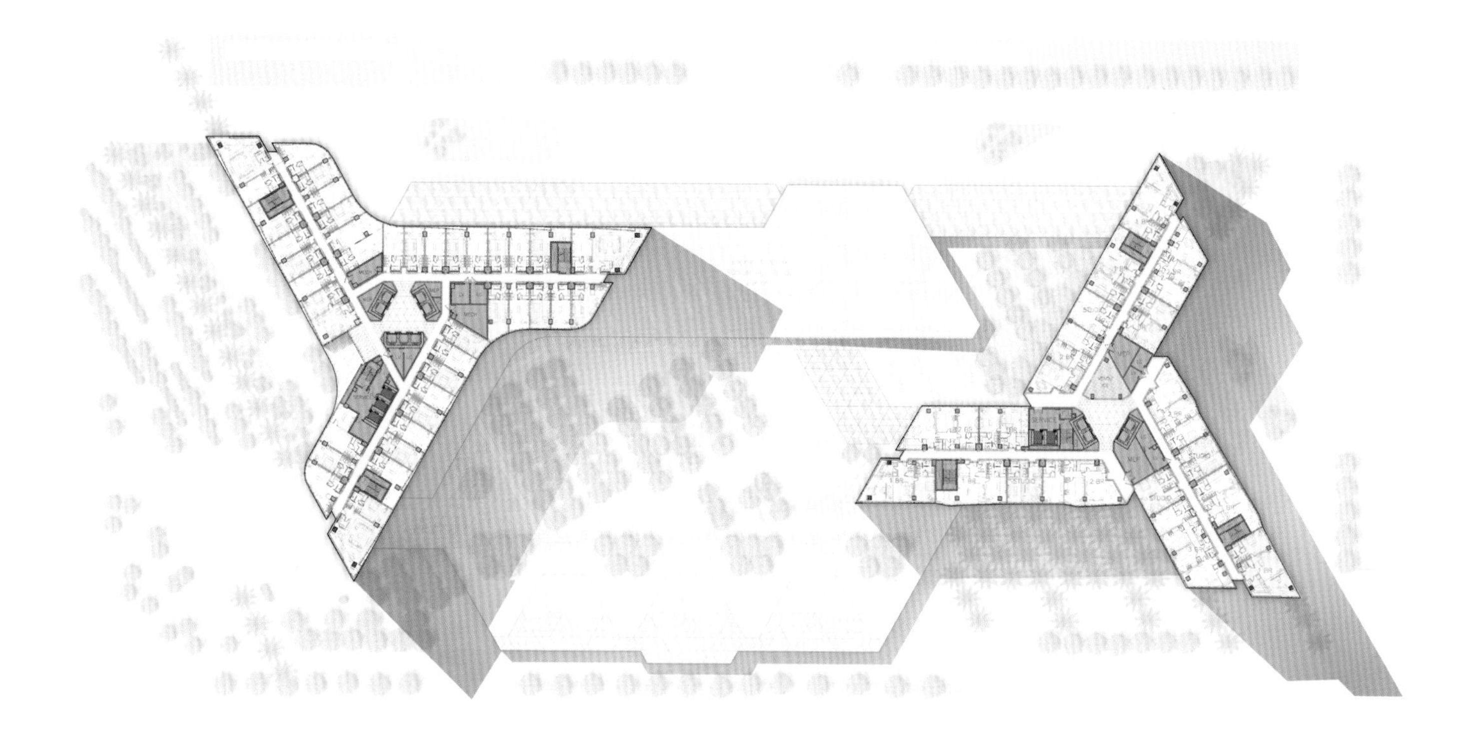

Hilton
Hilton
Hilton

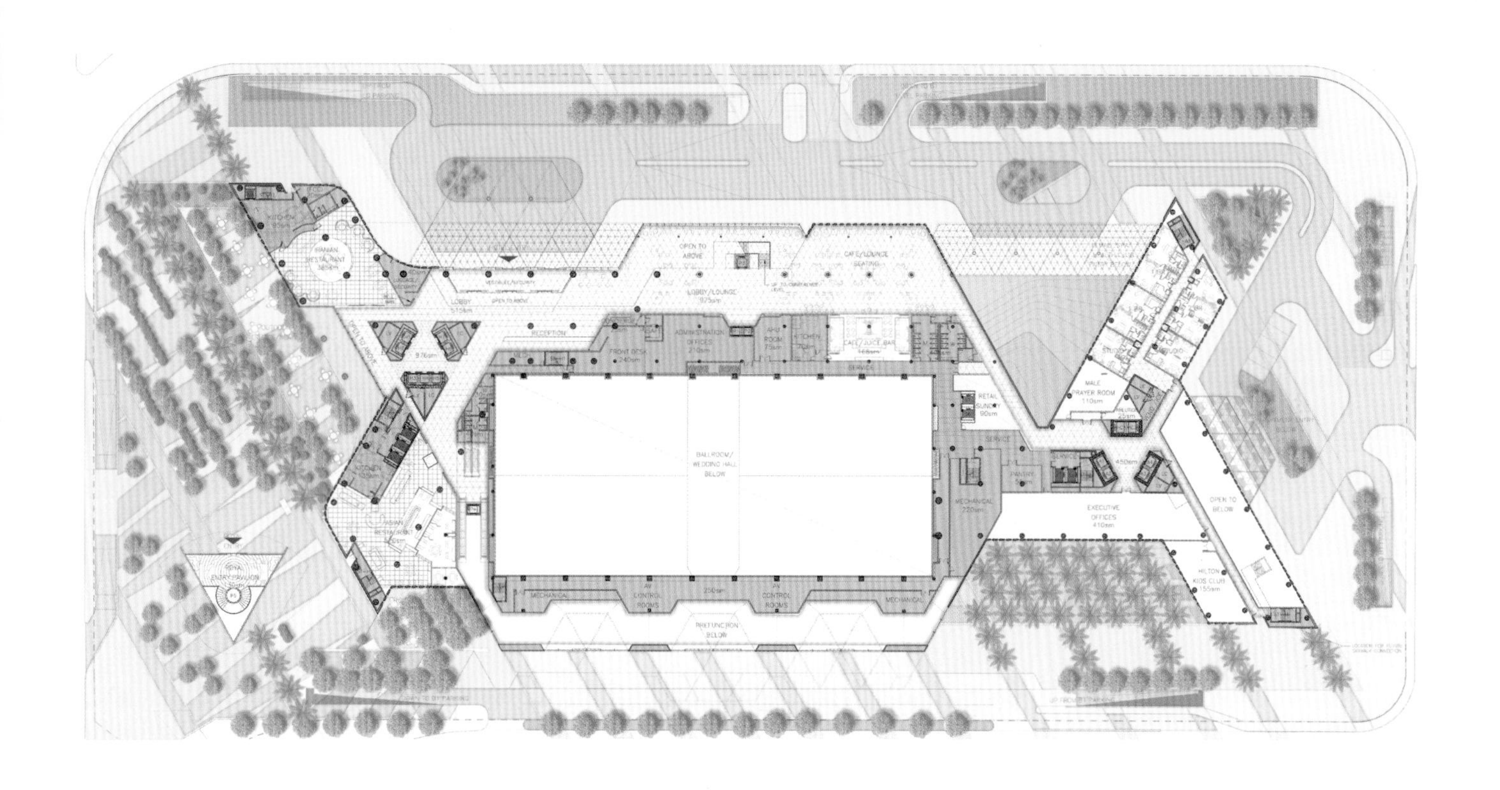
OPEN TO ABOVE
CAFE/LOUNGE SEATING
LOBBY/LOUNGE 975sm
LOBBY 515sm
RECEPTION
FRONT DESK 240sm
ADMINISTRATION OFFICES 210sm
AHU ROOM 75sm
CAFE/JUICE BAR 168sm
SERVICE
RETAIL 90sm
BALLROOM/ WEDDING HALL BELOW
MECHANICAL 220sm
AV CONTROL ROOMS
250sm
MECHANICAL
PREFUNCTION BELOW
ASIAN RESTAURANT

Hilton
Hilton
Hilton

门尼察遗产大厦
MENNICA LEGACY TOWER

Warsaw, Poland

The Mennica Legacy Tower complex, part of a newly approved master plan for development in the CBD of Warsaw, is divided into a 35-story tower on the southeast side of the site and a smaller 10-story building on the west side. The development features approximately 80,000 square meters of Class A office space, including a conference center, fitness center and ground-level retail, as well as four levels of underground parking, and ancillary services. A large plaza between the two structures provides ample space for outdoor seating and a variety of landscape features.

The taller building has a three-story lobby with a cable-supported enclosure, utilizing a low-iron glass with a non-reflective coating that blurs the boundary between interior and exterior space. The tower core is clad with large stone slabs that accentuate the solid mass of the core, in contrast to the lightness of the building's lobby.

The 2,000-square-meter tower floor plate provides an almost column-free space with 11- to 13-meter lease spans and a 1.35-meter planning grid. The tower mass features rounded corners on the northeast and southwest sides that help to reduce the visible length of the east and west façades; on the opposite corners, the tower incorporates a strong vertical edge.

The southeast and northwest corners are further highlighted by slightly recessed vertical slots that break up the building's mass and introduce a distinctive feature. The southeast corner of the tower steps in three-floor increments outward as it rises, opening the slot to the sky and creating a dynamic and unique profile that becomes a glowing beacon at night. A sloping screen wall at the building's top enhances the profile and integrates roof terraces into the overall massing. The textured, saw-toothed façades reinforce the rounded corners and give the enclosure an ever-changing appearance as one moves around the building.

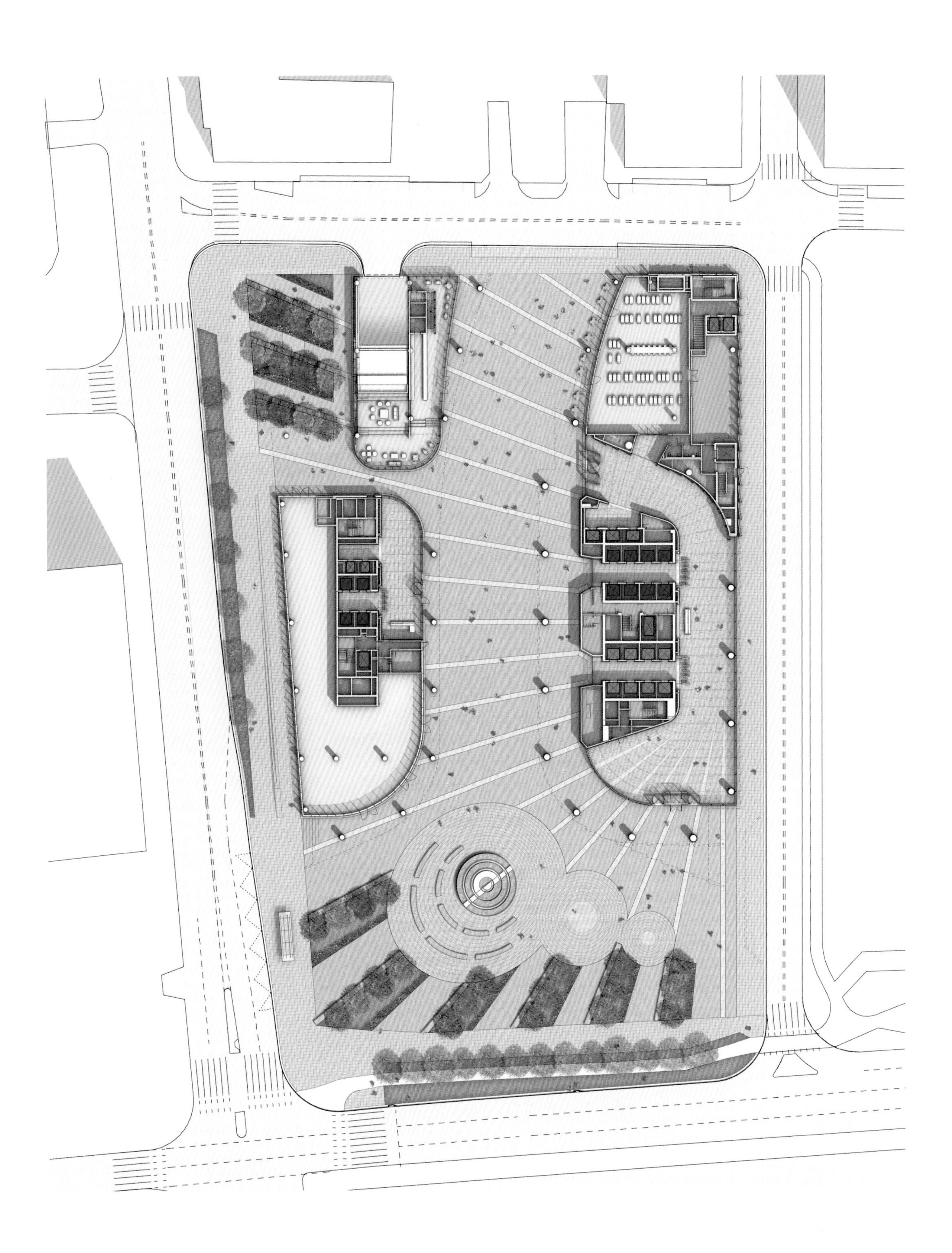

深圳800地标大厦
SHENZHEN 800 LANDMARK TOWER

Shenzhen, China

The proposed 800-meter mixed-use tower in Shenzhen is intended to create a global landmark for the fastest growing metropolis in China. The bold design announces the city's status as a technology and financial hub, while eclipsing the height of all other structures in neighboring Hong Kong, Macau, and Guangzhou. Located at the heart of the Luohu district, the project both physically and spiritually marks the center of Shenzhen, dating from its origins as an ancient village to the modern-day megacity.

The massing for the tower is composed of a series of "structural tubes" that taper and step in height as the building rises from the center of the famed Luohu mountain range. At their peaks, each tube is sculpted to create a blossoming effect, signifying the growth of the city as well as the broader region. The last occupiable floor is 740 meters in the air and will serve as the world's tallest observation deck when complete.

The 140-story tower is composed of Class A office space, a five-star hotel, luxury residential units, and boutique retail, in addition to the public observation deck. The overall massive development includes a large urban park and nearly 30 additional buildings, many of which will be residential and commercial projects that leverage the impact of the megatall tower.

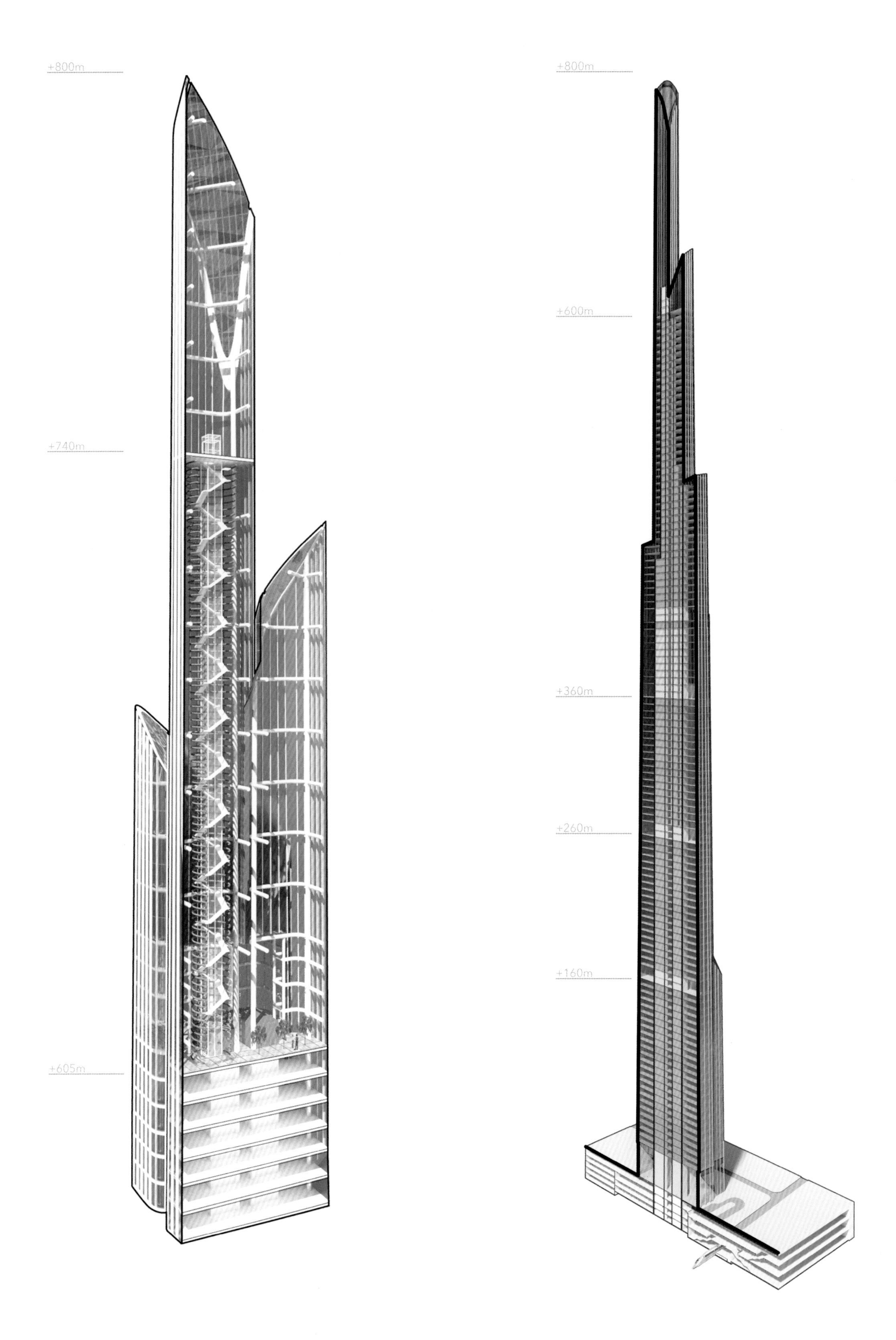
+800m
+740m
+605m
+800m
+600m
+360m
+260m
+160m

苏悦广场
THE SUMMIT

Suzhou, China

The Summit is a two-parcel development in Suzhou, China, that connects to the existing metro station and borders both sides of Suhua Road, the city's ceremonial boulevard and center for commercial development. The mixed-use development features two towers–one 16 stories and the other 39 stories–and includes office, retail, and luxury residences that offer unobstructed eastern views to Jinji Lake.

The design concept organizes the various program elements into a series of interlocking volumes. Each volume is sized to provide optimal functional depth for the program contained within, while creating a compositional quality that visually unifies the two parcels. An innovative gridded façade system is utilized for both towers to further visually connect the buildings while seamlessly integrating operable ventilation for all users. LEED Gold certification is mandated for the development, and direct connection to mass transit, extensive green roofs, locally sourced materials, and high-performance enclosures are a few examples of the sustainable strategies employed throughout.

The efficient, modern towers also strive to translate elements of context into the architectural expression. Therefore, the project color palette of white, grays, and black has been inspired by the architectural vernacular famous in Suzhou. Capturing an essence of the city is critical, as the project will enjoy unparalleled visibility in the downtown due to its location along the ceremonial boulevard and its immediate adjacency to the active public plazas to the east.

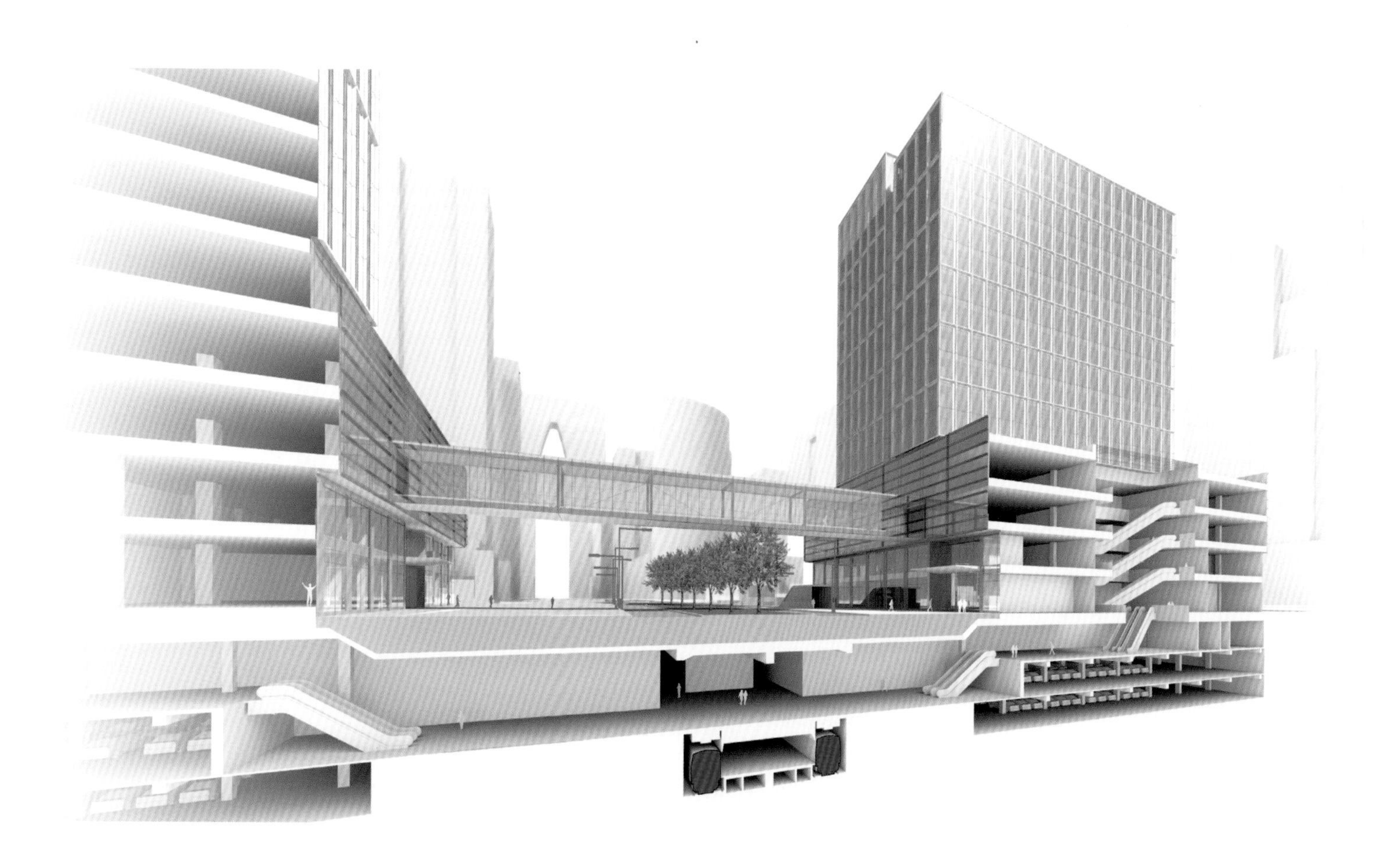

Salvatore Ferragamo

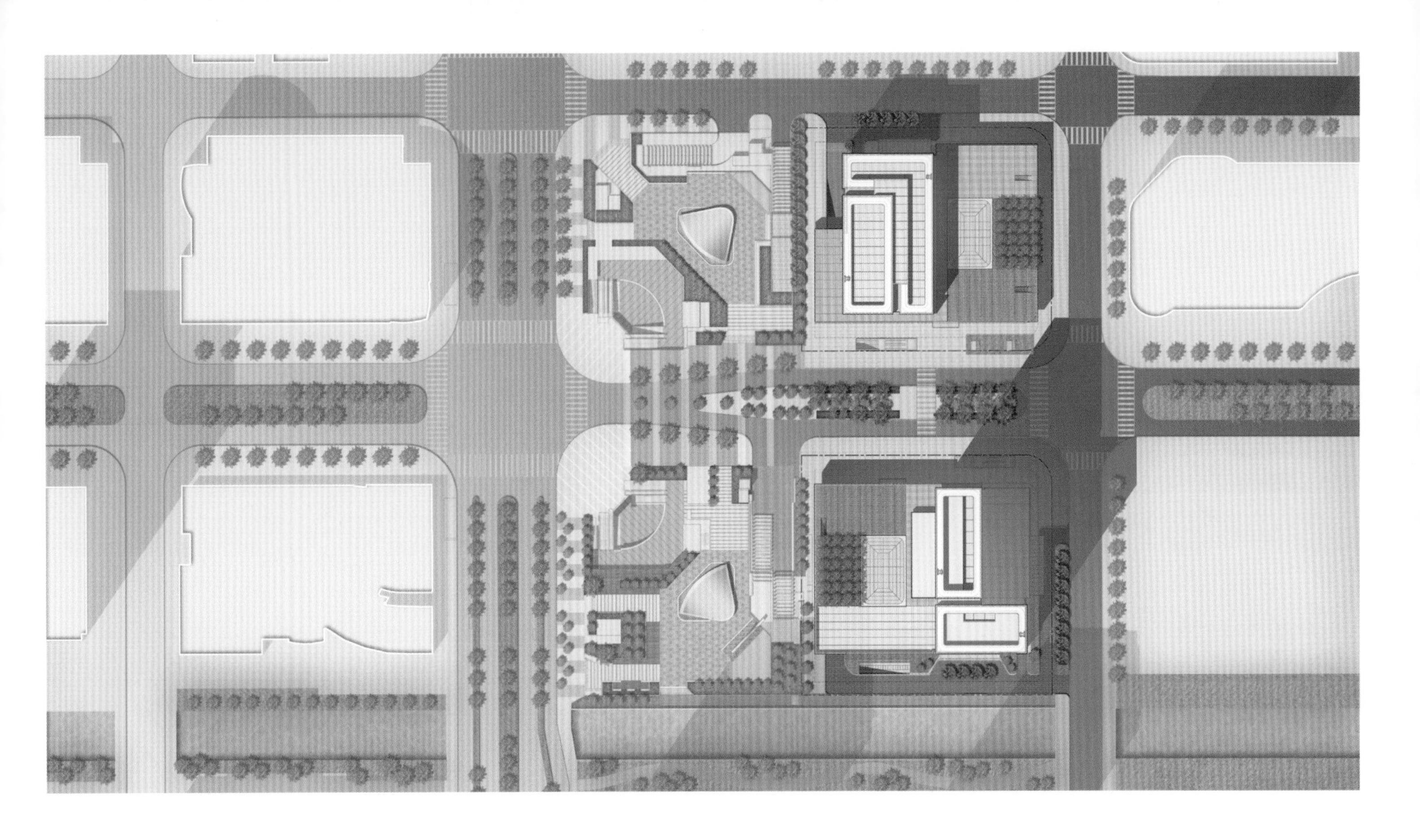

中英人寿保险公司前海大厦
AVIVA-COFCO QIANHAI TOWERS

Shenzhen, China

The two-tower Aviva-COFCO Qianhai complex is located in the rapidly developing Qianhai district of Shenzhen, China. The 200-meter COFCO Asia Pacific Tower will serve as the headquarters for COFCO's Asia Pacific group, while the 180-meter Aviva-COFCO Financial Tower will provide space for the Aviva-COFCO Life Insurance Company.

Nearly twins, each tower is composed of several slender bars that step back to reveal a series of internal atria and a sky terrace in the upper zones. The tower façades are defined by large, distinct frames and curtain walls that accentuate the buildings' verticality. The main façade on each tower is slightly angled to enhance occupant views and maximize the view corridor between the two towers. The high-performance curtain walls are designed to remediate Shenzhen's climate by providing external solar shading and natural ventilation, with operable windows hidden behind perforated metal panels to reduce any visual disturbance on the overall façade.

Connected directly to the city's extensive pedestrian and public transportation system, the COFCO site serves as a neighborhood hub, linking the green belt to the east with nearby office towers. Both exterior and interior paths allow public access, resulting in a unified urban experience between the neighborhood green spaces.

The upper zone of the east tower is reserved expressly for COFCO headquarters, while the upper zone of the west tower will accommodate multiple tenants. The interior is designed to allow for flexibility and provide ample social space to promote employee interaction and an open exchange of ideas. A large internal atrium connects the social zones on each floor, creating a continuous open space from the main lobby up to the sky terrace. At the top of each tower, the large 300-square-meter sky terrace offers a spectacular outdoor amenity space.

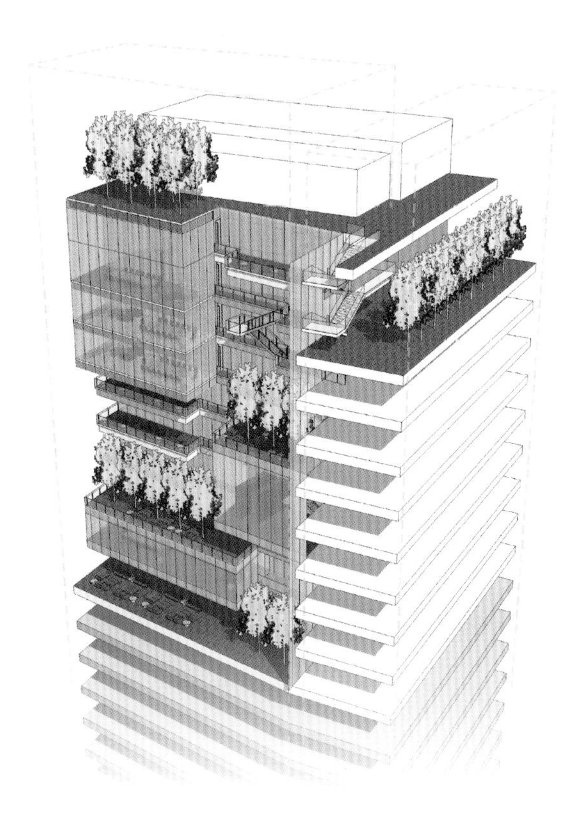

EAST TOWER
GAP

上海临港办公综合楼

SHANGHAI LINGANG INTERNATIONAL CONFERENCE CENTER

Shanghai, China

The Shanghai Lingang International Conference Center, in the heart of Harbor Front New City's downtown area, is by the East China Sea and the junction of the Yangtze River and Hangzhou Bay, adjacent to the Shanghai Harbor Front Service and Industrial Area and the Yangshan Free Trade Port Area. The master plan aims to create the most important "gateway" to Shanghai from the southeast and to provide access to the major coastal islands. The entire area is centered on the scenic Dishui Lake. The city street network expands outward in the form of rings and radii from the center of the round lake, creating a spatial characteristic reminiscent of water ripples. The programmed functions reinforce the idea of developing a modern, coastal, international city and a high-quality live-work environment.

The International Conference Center, in the northeastern quadrant of Shanghai Lingang, consists of four separate parcels each representing a city block. Each parcel is organized around central courtyards. Office buildings define and occupy each corner of the block, with a single-story retail component completing the courtyard ring.

The buildings are designed to be very simple in character and layout. Cores are situated at either end of the buildings so that the entire center is clear for efficient planning. Open office layouts have the potential to go from one exterior wall to the other, allowing natural light to penetrate from both sides. Most buildings have a terrace at the top floor, an outdoor space that overlooks the lake and other natural surroundings. There is a standard palette of glass and metal materials that is modified slightly between tall and short buildings and between the different parcels to create a lively development by using a few simple design features to distinguish individual buildings while also establishing continuity.

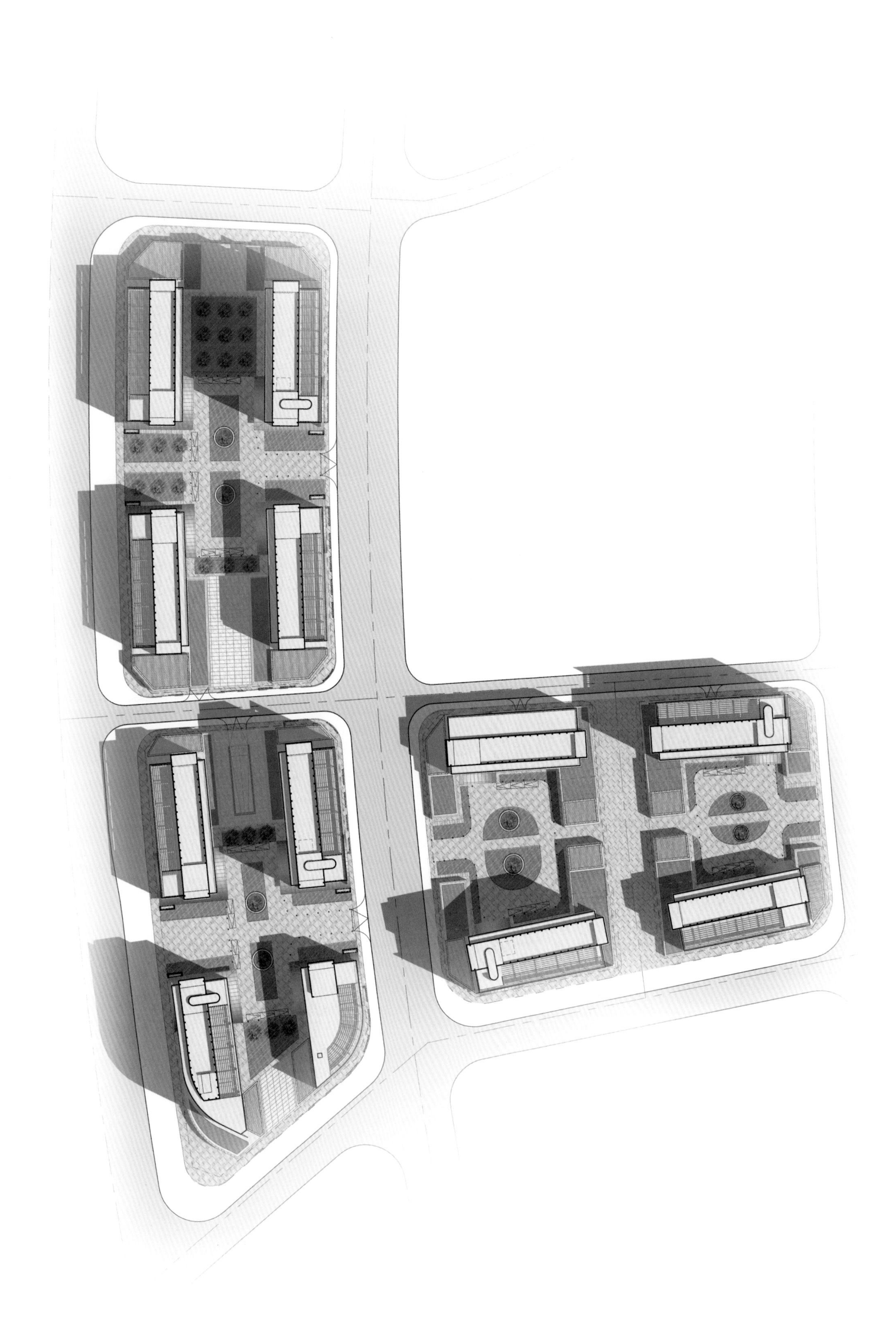

COFFEE FANTASY
DOMO CAFE

前海发展区
QIANHAI DEVELOPMENT

Shenzhen, China

The Qianhai district in Shenzhen is a special economic zone designated for an ambitious $45 billion development intended to transform the 15-square-kilometer area into the "Manhattan of the Pearl River Delta." This project, located in Neighborhood 2 of the district, covers 6.18 hectares and includes three office towers, a five-star hotel tower, and an apartment tower. The overall development totals more than 500,000 square meters and includes a shopping mall and retail stores in addition to the towers. GP is designing the five towers, as well as the hotel and apartment podiums and their affiliated retail store areas, and GP is collaborating with design firm Benoy, which developed the master plan and is designing the shopping mall.

The design concept emphasizes a unified complex composed of buildings with related yet individual exterior characteristics. The tower façades are designed with a textured elegance that differentiates them from the surrounding blue-glass buildings more common to Shenzhen. A silver-metallic-painted aluminum frame provides a sense of detail and enhances the tower's vertical appearance while catching and reflecting light. The frame also enhances the energy efficiency of the glass by shading it from intense daylight while providing even, filtered light to the spaces within. The spacing between horizontal frame elements varies from a two-story to a four-story rhythm to respond individually to each office building's height and proportions. The frame's vertical component is accentuated by means of double fins; this character is countered by an expression of double horizontal fins on the hotel and apartment towers that create a related yet different appearance while affording maximum flexibility for views and natural ventilation.

禅城绿地中心
CHANCHENG GREENLAND CENTER

Foshan, China

The 10-building Chancheng Greenland Center complex is sited in a prominent location in Foshan, China, along Jinhua Road, east of Wenhua Park.

The concept of balance is manifested in the overall layout of the program elements and in their relationship to one another. Initially, the project's towers were set within a simple grid responding to their general requirements: the Class A office towers with presence from the east and west along Jihua Road; the residential areas to the quieter and less visually prominent north end of the site; and the commercial office towers in between. The buildings were then shifted north, south, east, or west to ensure that all towers receive ample sunlight and offer pleasing views, while also maintaining their required minimum distances from one another.

The shifting of building locations is achieved in a push-pull manner, so that each tower is related to the next within the asymmetrical yet balanced site plan. The resulting organization, form, and linking of the towers ensures that the development feels like a single, unified place. The plan also defines secondary spaces that create the main internal circulation paths of both the street retail and shopping mall and become open spaces for pedestrian plazas, entry drop-off, and below-grade ramp access.

Green roofs, water pools, garden spaces, landscaped pedestrian plazas, and façade shading are some of the strategies by which the complex addresses the natural environment. An integral part of the project's design concept is to maintain the feeling of nature's presence and openness while promoting a network of accessible green spaces. These areas and pedestrian plazas within the body of the complex create a linkage with the nearby subway stations and ensure local pedestrian circulation is drawn into the complex.

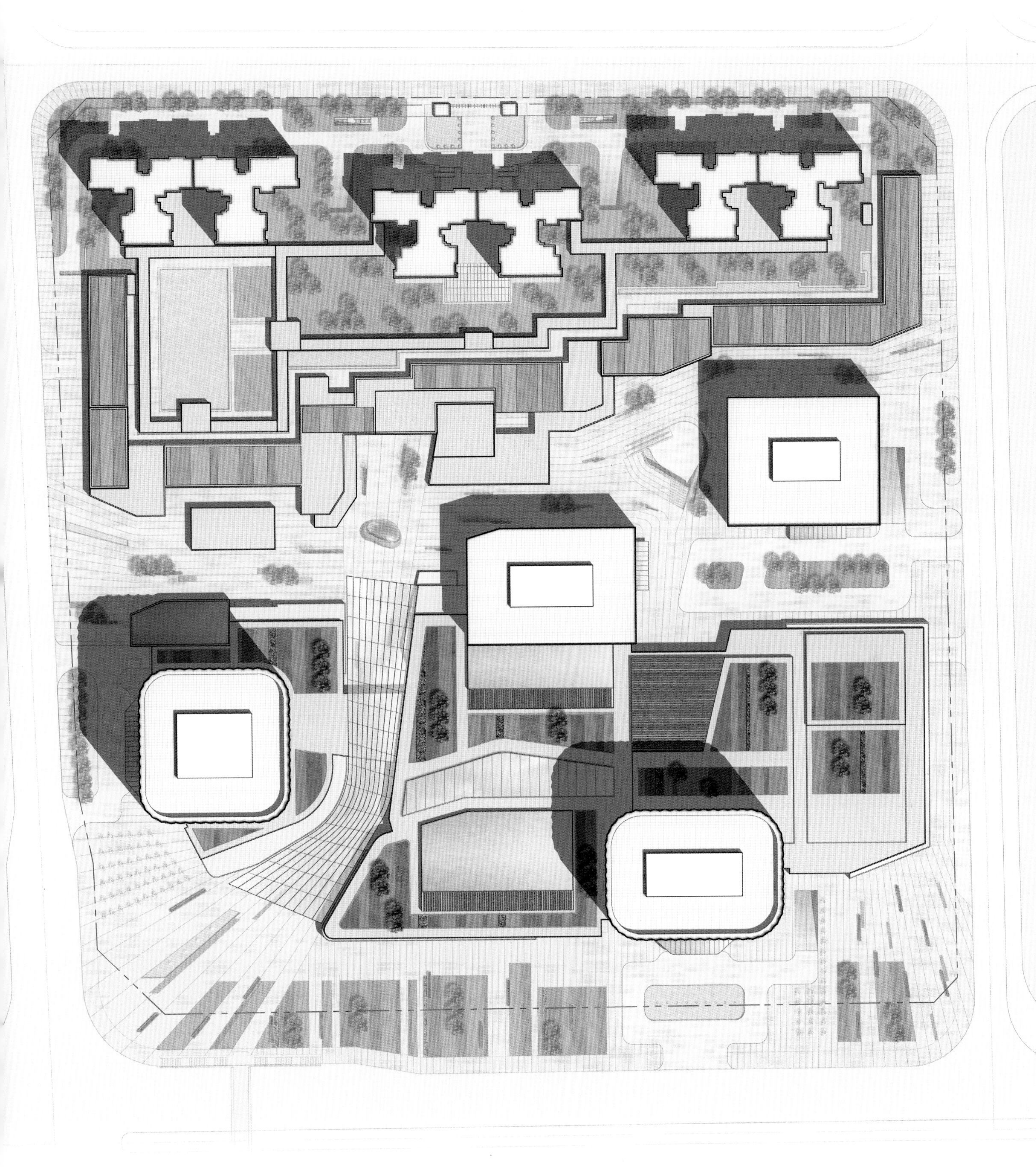

adidas
ZARA
GUCCI
HERMES

禅城绿地中心二期竞标

CHANGCHENG GREENLAND CENTER–PHASE 2 COMPETITION

Foshan, China

The conceptual approach for the project is to maximize the significance of the site's Chancheng District location by recognizing the nearby parks, stadium, and public transportation outlets as value-adding catalysts and formalizing a flowing path of landscaped open spaces or plazas between them. Buildings are then strategically arranged along the path to establish meaningful programmatic connections, take advantage of external adjacencies and views, and provide clear, separate pedestrian and vehicular circulation routes.

Retail is the project's lifeblood and recognized as the most essential ingredient for a vibrant complex. The centralized retail is located at the north end of the site to create a presence along Jihua Road. It also has proximity to both the subway and a transit station at the northeast and southeast corners respectively. The L-shaped neighborhood retail, adjacent to the residential buildings on the south side of the site, forms an urban edge while also maintaining permeability to attract people into the complex. The neighborhood retail is connected to the centralized retail by means of plazas and bridges. All retail benefits from direct connections to the subway and transit station at both the ground and B1 levels.

The landmark towers are meaningfully located on the site to work in tandem yet achieve singular objectives. The supertall Class A office tower is positioned at the northeast corner of the site with direct proximity to the subway. Together with an adjacent traditional office building, it forms a "gateway" entrance from the east and creates a strong presence toward the approach from Guangzhou. The hotel and apartment tower, located to the west of the site, uses its position and triangular shape to maximize views of Wenhua Park and its nearby public spaces. With its floating ballroom podium, this building provides the western gateway to the project.

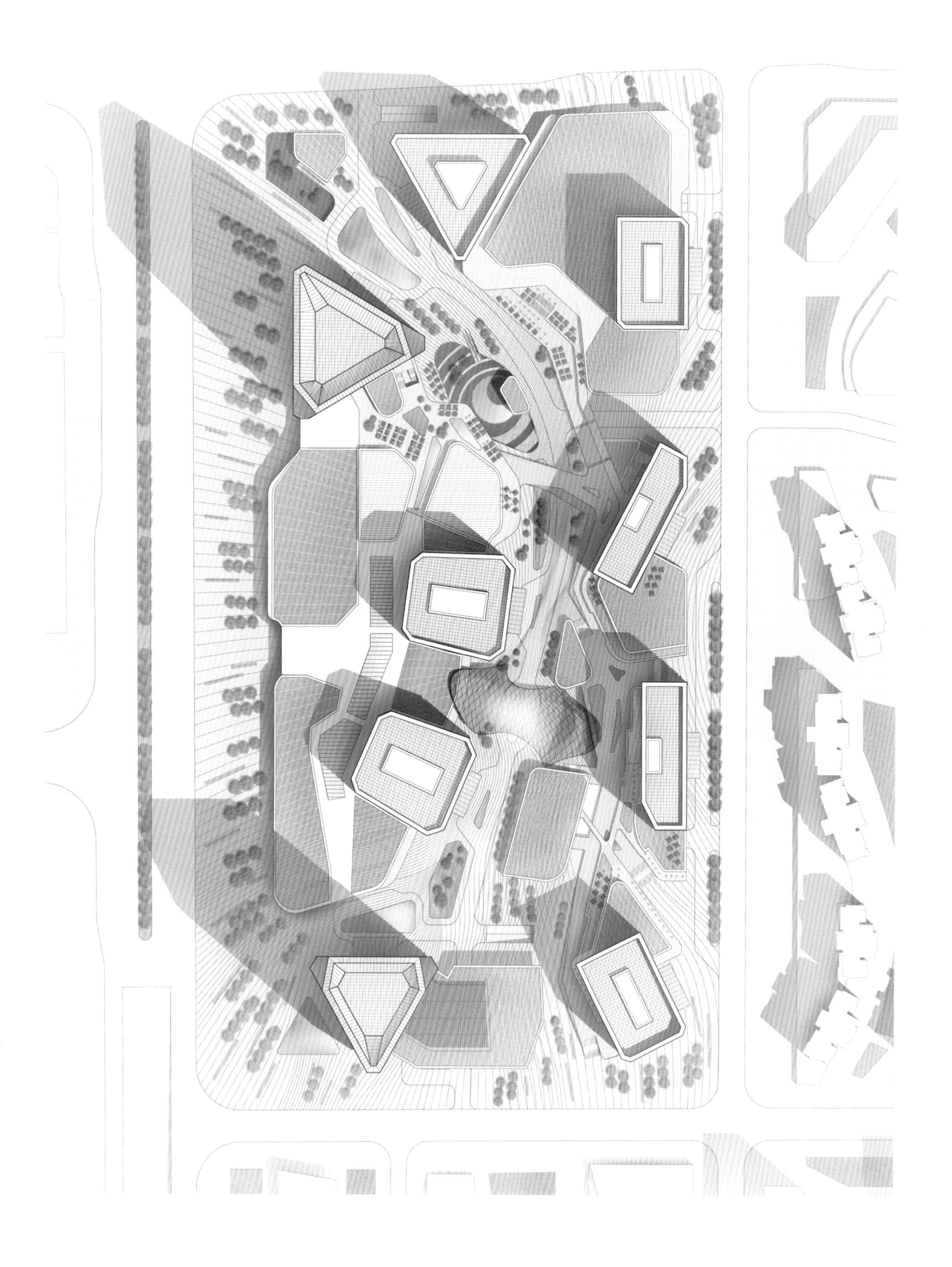

Dior
PRADA
Cartier
GUCCI
STARBUCKS

中国鸿荣源前海冠泽商业综合体

CHINA HOROY QIANHAI GUANZE MIXED-USE

Shenzhen, China

This mixed-use complex in the Qianhai district of Shenzhen integrates two Class A office towers, a five-star hotel, a retail center, and two serviced apartment buildings to create a dynamic, sustainable live-work-play experience that solidifies Shenzhen's significance in today's China. GP is leading the design of the office towers and the hotel.

The landmark for the development, a 62-story, 293-meter office tower, is designed as a simple, elegant structure anchored by four pairs of mega-columns marking the building's chamfered corners. As the building rises, the corners slope inward, with the tower culminating in a low-slung V shape to define the building top. A textured façade on the tower's four sides contrasts with the smooth, reflective corner façades, providing a distinguished presence in the skyline.

At 37 stories, the second office tower is designed to complement the landmark tower, featuring two similar chamfered corners and a comparable façade design. The floor plates in both towers offer flexibility to accommodate a variety of tenants. Optimized daylighting and spectacular views ensure a premier office environment.

Operated by Hilton as a Conrad property, the hotel includes 298 guestrooms and features cascading amenity floors and landscaped roof gardens along its western side, oriented to provide views of the adjacent green belt and Qianhai Bay. The guestrooms feature large floor-to-ceiling windows that take advantage of these views, and a variety of sophisticated amenity and dining options define the hotel as a prime destination.

The five-level retail complex connecting the office and residential towers acts as the engine of the development. The programming of the retail center is conceived to provide a new urban hub for Qianhai Bay.

The complex is designed with an emphasis on the outdoor environment, featuring an abundance of landscape, water features, and plazas throughout. With the podium levels lushly landscaped, the ground plane is effectively lifted to create a walkable setting with connections to the street level at specific locations.

CONRAD

武汉光谷绿地中心
GREENLAND OPTICS VALLEY CENTER

Wuhan, China

The Greenland Optics Valley Center complex, located in Wuhan, China, totals 315,000 square meters across two buildings. The project features a 400-meter office tower that will distinguish the Optics Valley along the skyline and expand the rich history of education and innovation for which Wuhan is famous.

The landmark tower is a fluid form combining rounded corners with a graceful taper that, together, aid in reducing wind pressures on the structure. A rhythmic pattern of exterior shades wraps the building while stretching and compressing to further articulate the tower base and crown. The use of color in the façade further reinforces the "digital" rhythm and provides a subtle reference to the local technology culture. The tower's height ensures unparalleled views of the neighboring mountain ranges, historic city center, and vast network of lakes and rivers that make Wuhan a prominent destination.

Upholding the developer's philosophy to "create better life," the project embraces sustainable initiatives in targeting LEED Gold certification. At the heart of the new district is a vast and meandering green belt that will function as the "spine" for the entire development. This large public amenity will connect the vast network of programmatic functions lining its edges, while ensuring that abundant natural light and generous parkland will be available for all structures. Integral to the environmental vision for the area, two new subway lines will have direct access to the parcels. Extending from the existing network, the Optics Valley will be a closely tied development. The new commuter node will become the epicenter of activity and ensure immediate integration into the broader city.

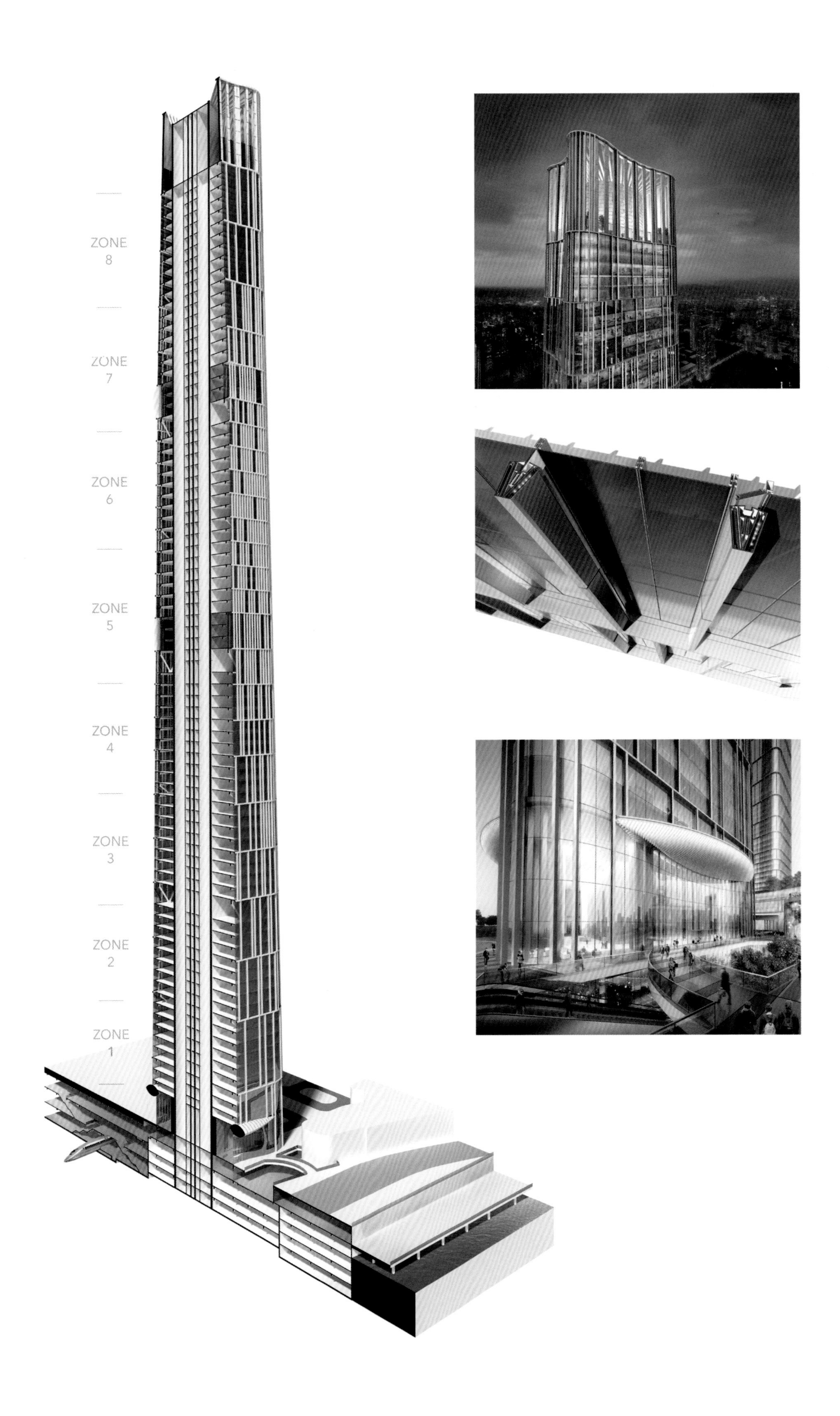
ZONE 8
ZONE 7
ZONE 6
ZONE 5
ZONE 4
ZONE 3
ZONE 2
ZONE 1

南宁华润中心

NANNING CHINA RESOURCES CENTER TOWER

Nanning, China

Located in Nanning, China, the capital of the Guangxi Province, the 403-meter Nanning China Resources Center Tower is sited along Minzu Avenue at the heart of the burgeoning Fengling District. The mixed-use tower is linked to public transportation through underground connections at the B1 level, and to adjacent buildings through indoor and outdoor pedestrian corridors at the ground and sixth floors respectively. Upon completion, the tower will be the tallest building in Nanning.

The design of the tower is derived from its multiple uses, which include more than 170,000 square meters of Class A office space, nearly 6,000 square meters of retail, and a 336-key Shangri-La hotel. The massing of the building steps and tapers to accommodate the changing floor plates of the various program types, resulting in a form that is both efficient and identifiable.

The angled geometries of the façades are designed to reinforce the crystalline form while celebrating the tower's verticality. Entirely encased in floor-to-ceiling, high-performance glass, the exterior enclosure features integrated ceramic shading elements that offer added solar control while maintaining ample natural light without obstructing views. Designed to LEED-CS Gold standards, the high-performance façade is one of many features holistically integrated toward reducing the project's environmental footprint while providing a world-class level of comfort and quality.

j'adore
Dior
GUCCI

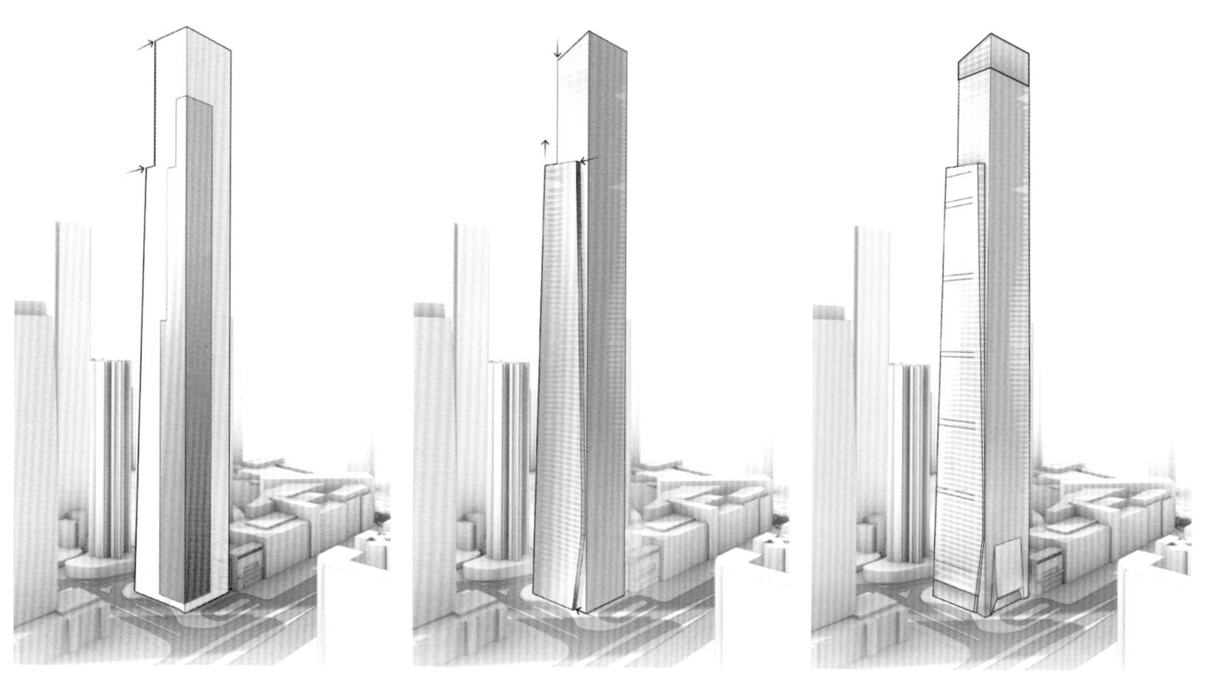

TOWER MASS TAPERS IN DIRECT RESPONSE TO THE STEPPING OF THE CORE

TWO INTERLOCKING VOLUMES CULMINATE IN ANGLED SCREENWALLS TO ACHIEVE A CRYSTALLINE EXPRESSION

MONUMENTAL SPACES AT THE BASE AND CROWN ARE UNIQUELY CLAD IN HIGH TRANSPARENCY FAÇADES

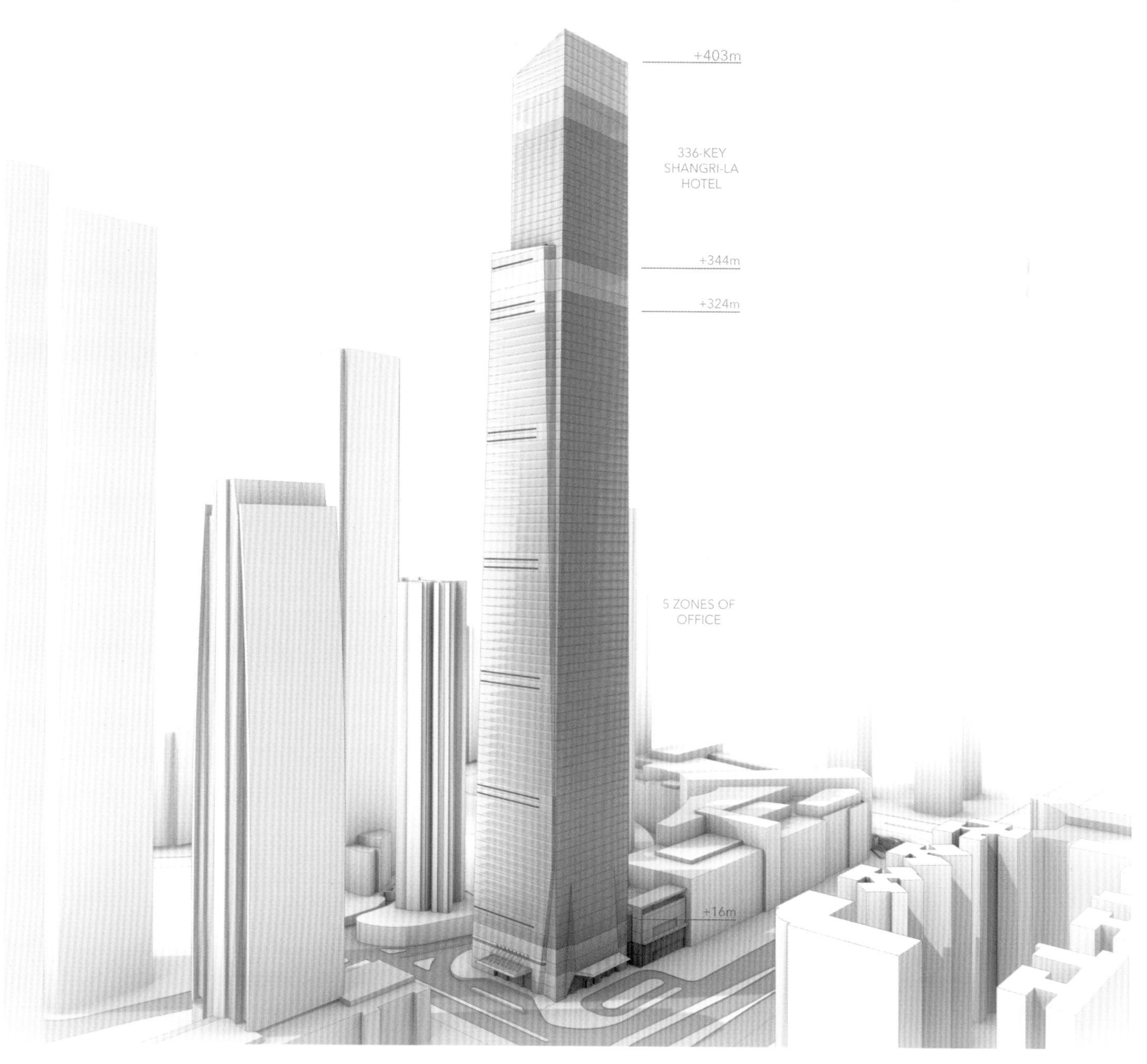

18
5
18
5
42
32

三亚保利瑰丽酒店与三亚广场

ROSEWOOD SANYA AND SANYA FORUM

Sanya, China

Located at the southern end of Hainan Island in the city of Sanya, China, the complex focuses on two conceptual design elements: a "lighthouse" for the hotel and serviced apartment tower, and a complementary "rock formation" for the convention center.

The 46-story hotel and serviced apartment tower rises from a conceptual outcropping of rock to become a landmark symbol of Haitang Bay, its glowing beacon visible from any direction. A unique resort in the sky, the building is organized vertically, with the arrival sequence, landscape, and incorporation of water and views all considered integral to the guest experience. The amenity-rich resort is purposely designed to be intimate and exclusive, offering 246 generously sized rooms—all with ocean views and individual terraces, some with their own plunge pool. The 465 serviced apartments are located in the main tower, which sits immediately behind the low, linear bar of hotel rooms. The top of the tower features a multilevel club for the exclusive use of the apartment residents and culminates in the "lighthouse beacon."

The Sanya Forum is designed as a premier convention facility, catering to national and international economic and political summits, as well as a variety of other meetings and special events. The building's rock-like mass is designed to be bold yet approachable. Once inside, a unique volume of space opens across multiple levels, with areas for meetings and formal ceremonies. The multifunctional building accommodates large events with an exhibition hall, ballroom, and forum space that can serve formal state gatherings, as well as private weddings and parties. The building also provides a variety of other meeting spaces, along with a large VIP meeting, gathering, and reception area at the top with an outdoor terrace overlooking the ocean. The design culminates in the private rooftop garden, with exclusive spaces juxtaposed against expansive ocean views.

改建项目
LEGACY

芝加哥伟基河畔南111号办公楼
111 SOUTH WACKER

Chicago, Illinois, USA

111 South Wacker is designed as a Class A multi-tenant office tower. The building's office space is positioned above seven floors of structured parking. The typical office floors are column-free, with 50- and 60-foot lease spans between the center core and the perimeter columns, optimizing efficiency and flexibility.

While the footprint of the building's upper floors covers almost the entire site, the building is open and spacious at street level. To create this sense of space, the perimeter tower columns transfer through the parking floors to provide large 80-foot spans at the base. The openness of the lobby is dramatically enhanced by the cable-supported, non-reflective glass enclosure and a very compact core. The result is the feeling that the ground floor is a covered outdoor plaza. Floor and ceiling patterns reflect the parking ramp above, creating a consistent, dynamic rhythm that energizes the space.

The enclosure systems are used to express structure and technology, the essential components of today's high-rise construction. The building is enclosed with a unitized curtain wall system with high-performance glass and featuring stainless steel V-shaped mullions that reinforce the vertical expression of the column cladding.

The building was designed from the outset with an emphasis on sustainability and LEED certification in mind. As a result of the sustainable design initiatives, the building became the first ever to be certified LEED-CS Gold.

...111 S. Wacker is a muscular skyscraper that reveals its internal structure rather than concealing it...

An oval-shaped lobby that slips beneath the building's boxy office and parking garage floors adds to the sense of spaciousness. It is wrapped in an extraordinarily transparent wall of cable-supported glass, almost making the distinction between inside and outside disappear...

–Blair Kamin, Architecture Critic, *Chicago Tribune*
"The Loop Gets Bold, Inviting Gateway," July 10, 2005

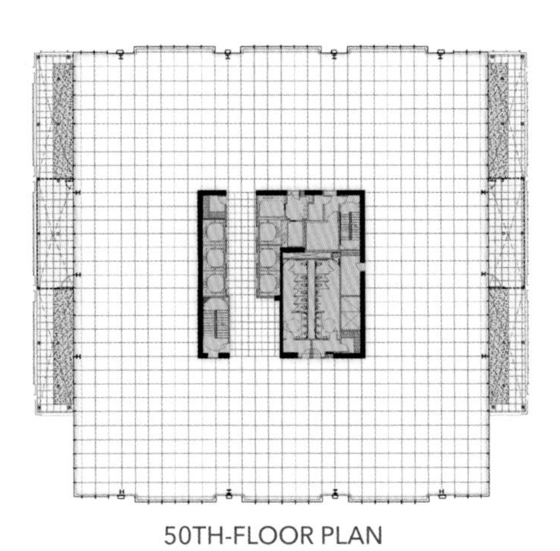
50TH-FLOOR PLAN

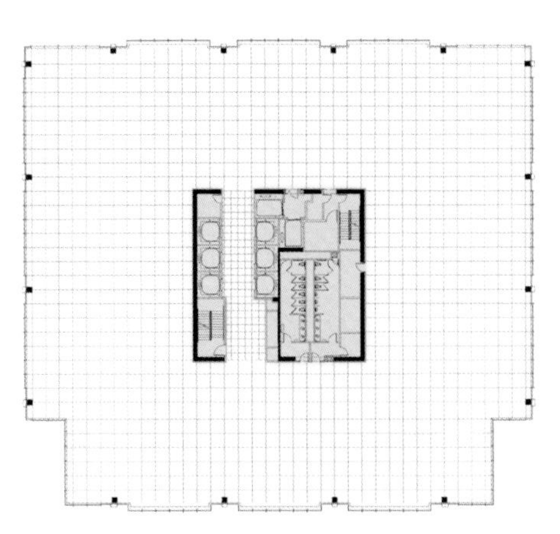
TYPICAL HIGH-RISE PLAN

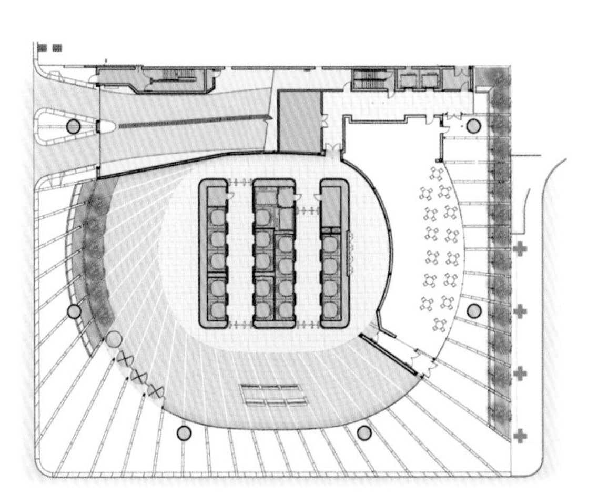
GROUND-FLOOR PLAN

鲁道夫街300号大厦
300 EAST RANDOLPH

Chicago, Illinois, USA

The 300 East Randolph building is a two-phased, vertically expanded office tower located in downtown Chicago at the north end of Grant Park. The facility primarily serves as the headquarters for Health Care Service Corporation (HCSC) and its Blue Cross and Blue Shield of Illinois division.

After relocating four times over a 20-year period due to continuous growth, HCSC realized it needed a plan to address the company's immediate space needs as well as accommodate anticipated future growth–without another move. The design team responded with an office tower that could be expanded vertically in phases.

Phase one provides 1,430,000 square feet of space across 33 stories, designed to accommodate 4,500 people in an open-office environment, along with a conference and training center and a 900-seat cafeteria. In phase one, the initial foundations and structure were planned, designed, and constructed to support the fully expanded building; additional riser space was also provided in order to accommodate independent mechanical, electrical, and plumbing systems for the expansion floors.

Phase two, finished more than a decade after phase one, vertically completes the building with 24 additional stories and 920,000 square feet of space, including a mid-building conference center added on level 30, providing necessary additional meeting and training space; a satellite cafeteria on levels 41 and 42; a fitness and wellness center; various conference and training facilities; a central chilled-water storage and distribution plant; and a green roof.

The fundamental mechanism for the expansion is an atrium along the north side of the building composed of five 40- by 30-foot open structural bays. Each of the two outer bays holds an eight-car elevator bank to service phase one. In phase two, an additional eight-car elevator bank is located in each of the two inner bays for the vertical expansion. The center bay is reserved for an open stair to facilitate inter-floor circulation, and every three floors, the center bay is fully built out and utilized for meeting space. This unique center-bay configuration also helps establish both a visual and physical connectivity within the company.

I wouldn't have thought you could do this. It's a great strategy to maximize site utilization, but somebody had to go first to prove that you could do it this way. They did.

–Juror, AIA Chicago Distinguished Building Awards 2011

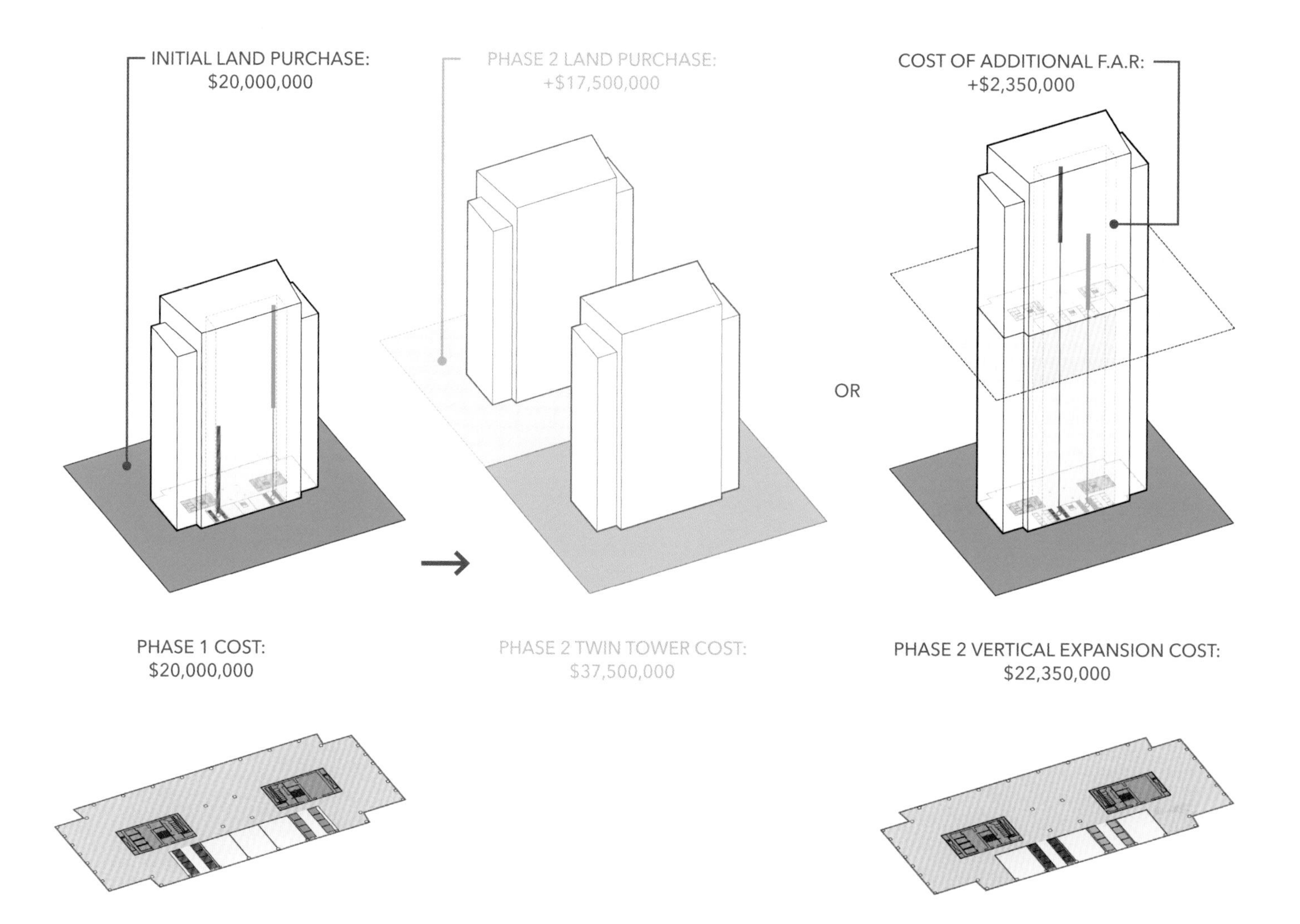
INITIAL LAND PURCHASE:
$20,000,000
PHASE 2 LAND PURCHASE:
+$17,500,000
COST OF ADDITIONAL F.A.R:
+$2,350,000
OR
PHASE 1 COST:
$20,000,000
PHASE 2 TWIN TOWER COST:
$37,500,000
PHASE 2 VERTICAL EXPANSION COST:
$22,350,000

瑞士银行大厦
UBS TOWER

Chicago, Illinois, USA

Located along the western edge of Chicago's Loop, UBS Tower is a Class A office high-rise with technical features that set it apart, at the time, from former generations of downtown office buildings. The 51-story tower is designed with three distinct floor plates of 29,300, 33,300, and 37,900 square feet and 45-foot column-free lease spans to maximize efficiency and flexibility for a range of tenants. The lobby is enclosed with a dramatic, 40-foot-high cable-supported netwall that uses low-iron glass with a non-reflective coating to create a virtually transparent boundary between inside and out.

The tower is enclosed with a unitized curtain wall with high-performance glass and linen-finish stainless steel mullions. Circular steel columns extend the building's full height and above the roofline, providing a distinctive image on the skyline and setting a clear organizational rhythm for the building's architecture.

Locating the tower in the northwest corner of the site allows the building to define the Wacker Drive street wall. This location also opens up the east side to a large plaza with fountains and creates a major pedestrian connection along Madison Street. This east-west connection is enlivened with large granite benches encircling mounded landscape elements and tall trees. Together, these elements reinforce the building's design and provide seating and shade for the thousands of pedestrians who traverse the plaza daily.

We love the way the ground plane is both inside and outside. This is a strong gesture that unifies the site plan. It's just beautiful in its composition, both simple and social.

–Jury, ASLA Professional Awards 2007

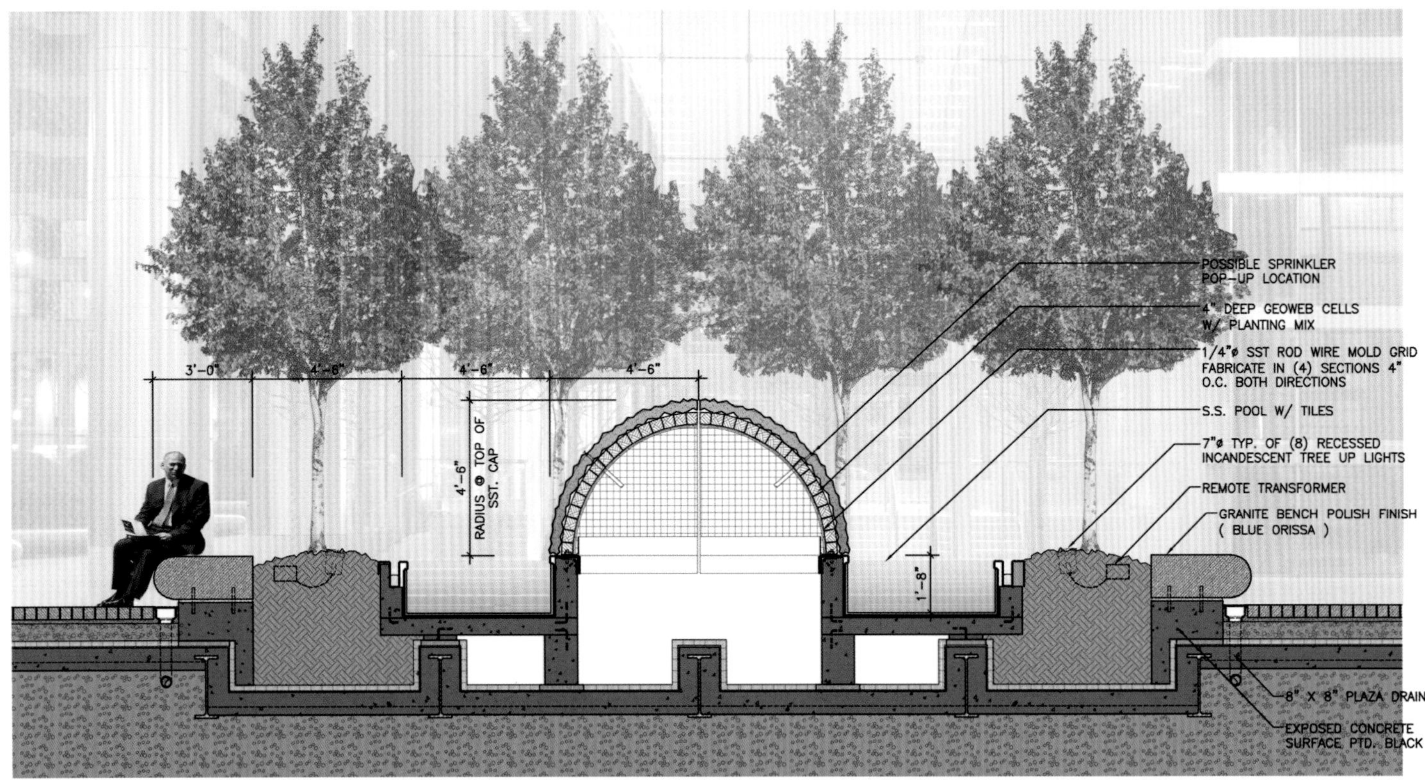
POSSIBLE SPRINKLER
POP-UP LOCATION
4" DEEP GEOWEB CELLS
W/ PLANTING MIX
1/4"ø SST ROD WIRE MOLD GRID
FABRICATE IN (4) SECTIONS 4"
O.C. BOTH DIRECTIONS
S.S. POOL W/ TILES
7"ø TYP. OF (8) RECESSED
INCANDESCENT TREE UP LIGHTS
REMOTE TRANSFORMER
GRANITE BENCH POLISH FINISH
(BLUE ORISSA)
3'-0"
4'-6"
4'-6"
4'-6"
4'-6"
RADIUS @ TOP OF
SST. CAP
1'-8"
8" X 8" PLAZA DRAIN
EXPOSED CONCRETE
SURFACE PTD. BLACK

LYRIC OPERA
OF CHICAGO

LYRIC OPERA

查尔斯广场中心
CHARLES SQUARE CENTER

Prague, Czech Republic

This building is located on historic Charles Square, one of the oldest and largest public spaces in Prague. The developer's goal was to provide a state-of-the-art, American-style office building in the historic city center. The eight-story building was constructed above an existing metro station and includes 15,300 square meters of Class A office space, 5,700 square meters of retail, and below-grade parking for 150 cars.

Prague is a city that is uniquely conscious of its history, and the opportunity to construct a new building there is rare, as the demolition of existing buildings is seldom permitted. The site for this building was bombed during World War II and had not been built upon since. The primary challenge was to design a contemporary office building that fulfilled today's business requirements for flexibility and technology, while respecting Prague's historic architectural context. The massing of the building recognizes the street wall and cornice lines of adjacent buildings. Fritted glass and embossed aluminum curtain wall panels provide the facade with a sense of scale and detail that complements the architectural traditions of Prague.

The first two floors are occupied by retail tenants that benefit from a skylit atrium and foot traffic from the metro station below. The atrium is reminiscent of the traditional Prague courtyard building, as the hard-edged street wall gives way to an interior moment of light and air. The typical office floor plates provide planning efficiencies and state-of-the-art building systems that are rare within the historic city core.

It's a refined contemporary expression that integrates well into the historic fabric of Prague.

–Michael Newman, President & CEO, Golub & Company

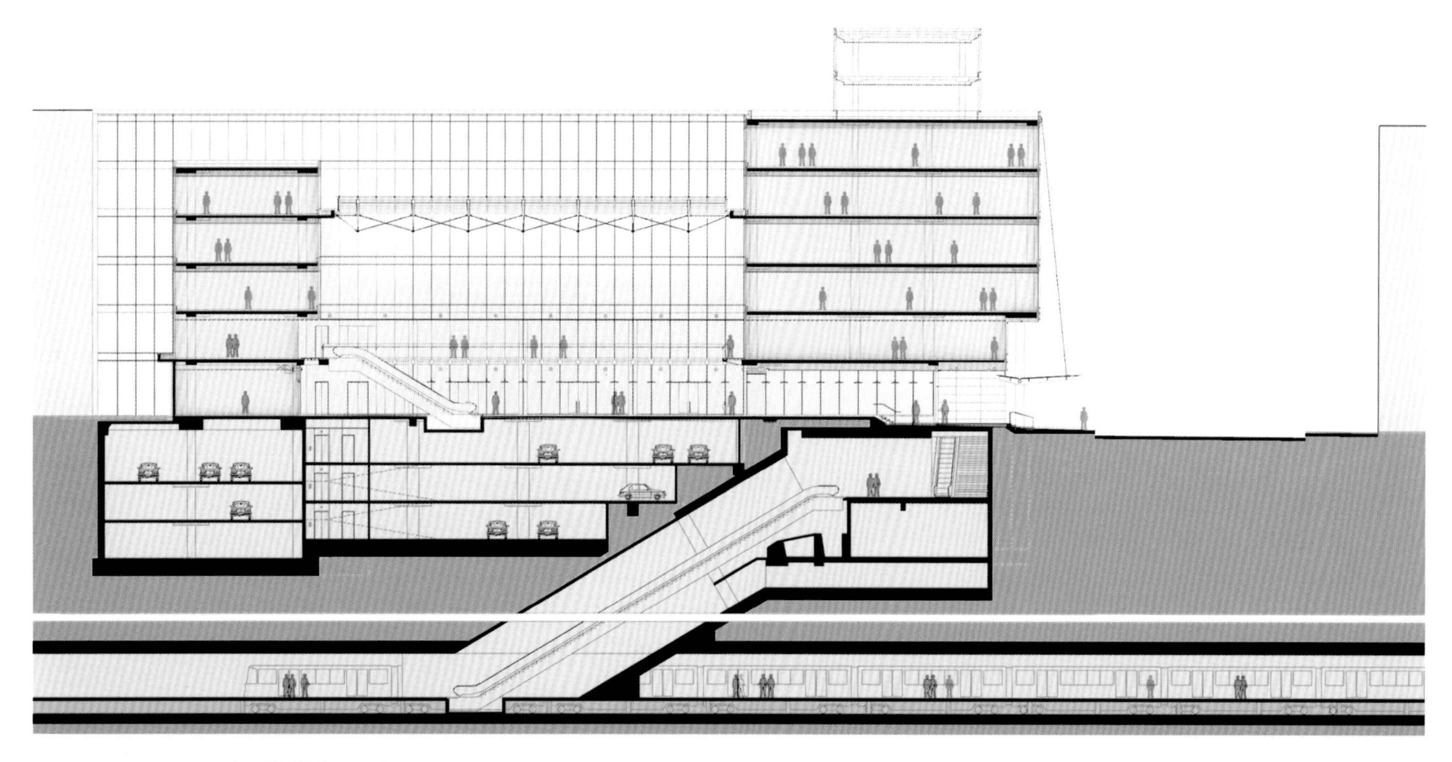

Raiffeisenbank

NESCAFÉ
NESCAFÉ

中国钻石交易中心大厦
CHINA DIAMOND EXCHANGE CENTER

Shanghai, China

The China Diamond Exchange Center is a 50,000-square-meter office building in Shanghai's Pudong district, the city's modern financial and commercial hub. The 15-story building provides space for members of the China Diamond Exchange, as well as other related tenants.

The building is conceived as two rectangular office slabs connected by a skylit atrium, with a large 66- by 230-foot cable-supported netwall at each end. One tower is fully dedicated to the Diamond Exchange members, with the adjacent tower serving other tenants. The relatively narrow 20-meter-wide floor plate of each tower is unique, working in combination with the atrium to bring in daylight throughout, as natural light is preferred by the diamond trades.

The open elevator tower is the focal point of the building, with three cabs traversing the atrium to sky bridges on each level that connect the two office blocks. The elevators' activity is not only visible from the lobby but also from outside through the full-height glass walls.

The major tenants' core business inspired the design, with diamond-shaped elements featured in key areas, including the atrium's glass skylight, the structural geometry of the entry canopy, and the main lobby floor. Details provide punches of red, a color that signifies prosperity and happiness in Chinese culture.

With its signature atrium, the building resembles a polished gem, symbolizing the nature of the business conducted within.

—Peter Tao, President, Shanghai Lujiazui Development Co.

圣保罗君悦酒店
GRAND HYATT SÃO PAULO

São Paulo, Brazil

Sited between two skyscrapers in the Marginal Pinheiros business and financial district, the Grand Hyatt São Paulo offers spectacular views of the city and the Tietê River. The five-star hotel maximizes its relatively dense urban site, providing 466 guestrooms, a two-story dining and entertainment complex, a convention facility with 3,000 square meters of meeting space, and a fitness center and spa. The facility also includes a business center and on-site parking for 600 cars.

In its downtown location, the complex caters to international business travelers with three distinct elements: a hotel tower, a conference center, and a separate dining facility that also serves residents of São Paulo. The hotel is a crisply sculpted tower of stone, glass, and steel that addresses its corporate neighbors and provides a distinct presence on the skyline. Two efficient double-loaded corridors, which are linked to a central elevator core, provide visual access to landscaped gardens and pools that lend a resort-like quality to the otherwise all-business hotel.

The glass-enclosed lobby is 10 meters high and features granite and native orange-hued Louro Faia wood. A glass bridge and exposed elevator connect the fitness center and spa to a large outdoor pool. The conference center is housed in a connected yet clearly distinct structure and has two ballrooms and seven meeting rooms that offer natural daylight and private terraces. The exterior details of the restaurant complex mirror those of the other structures; however, its gray-colored stone differentiates it from the yellow hue of the hotel and conference center.

Entered via a three-story, light-filled lobby, the Grand Hyatt…is indeed grand, but also surprisingly serene.

–Hot List, *Condé Nast Traveler*, May 2003

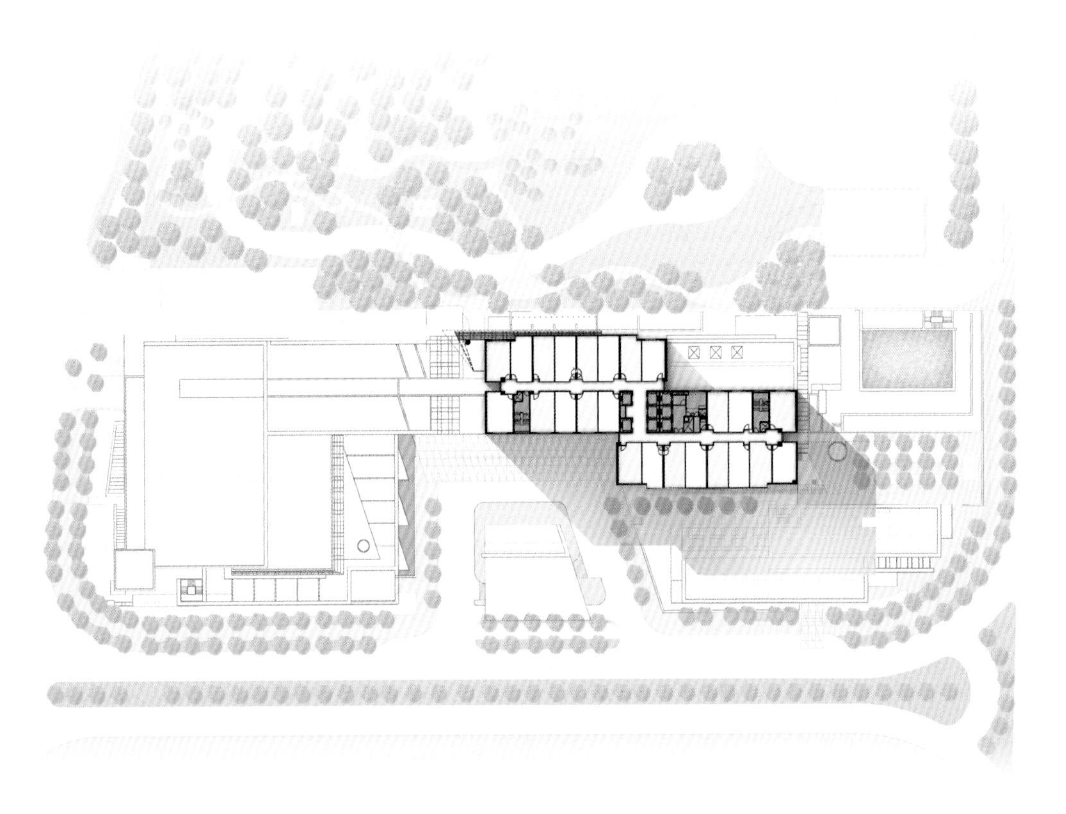

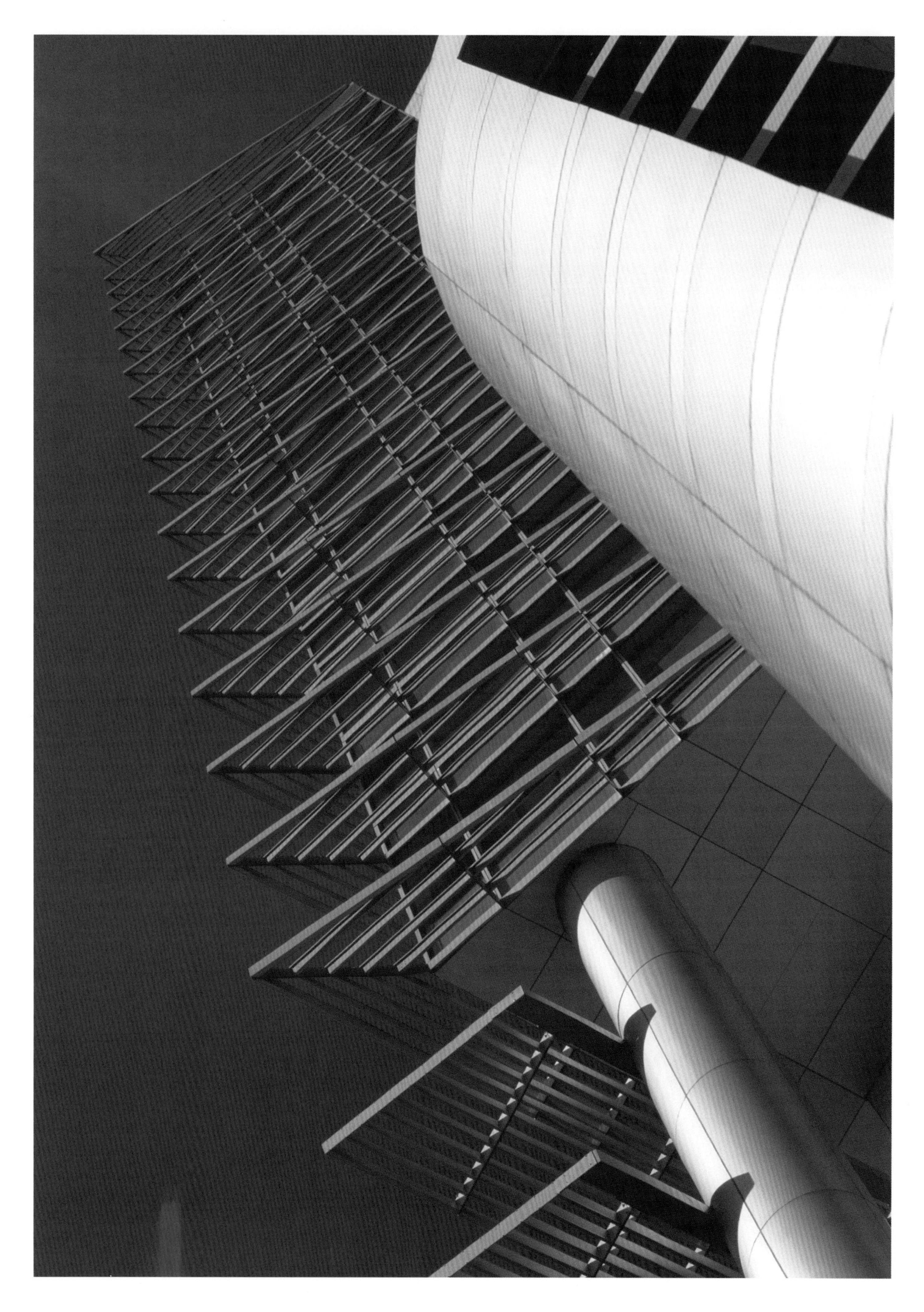

孟买君悦酒店
GRAND HYATT MUMBAI

Mumbai, India

This luxury hotel and residential development sits on 5 hectares located just 20 minutes south of Mumbai's international airport. The five-star property includes 547 guestrooms, 110 serviced apartments, a business center, health spa, fitness center, pools, and 10,000 square meters of retail. In addition to the pools, outdoor recreational facilities include basketball, tennis, and volleyball courts, as well as a paved jogging path. The property is distinguished by stunning water features, distinctive hardscape elements, and lush landscaping, plus many sculptures and other works of art throughout.

Due to its proximity to the airport, the structure could not be more than six stories tall, and local mandates required that 15 percent of the site remain open. To attract major conferences and India's customary weeklong wedding celebrations, the hotel includes a 1,000-seat grand ballroom, meeting rooms, four restaurants, an entertainment center, and below-grade parking for 700 cars.

Challenging zoning requirements called for an ambitious design to meet the expectations of sophisticated international travelers. The courtyard complex, with a large, landscaped garden in the middle, is modern yet timeless, with subtle local and cultural references. Each component complements the others while retaining a distinct identity and a clear separate entrance. The main building, with its deep porte-cochère, has a curved façade of banded glass and aluminum that stands in subtle contrast to the stone-clad wings.

The hotel is an oasis and true design inspiration in the heart of Mumbai.

—Kurt Straub, Vice President—Operations (India),
Hyatt Hotels & Resorts

NO DIVING
NO DIVING
4'3"
1.30 mts

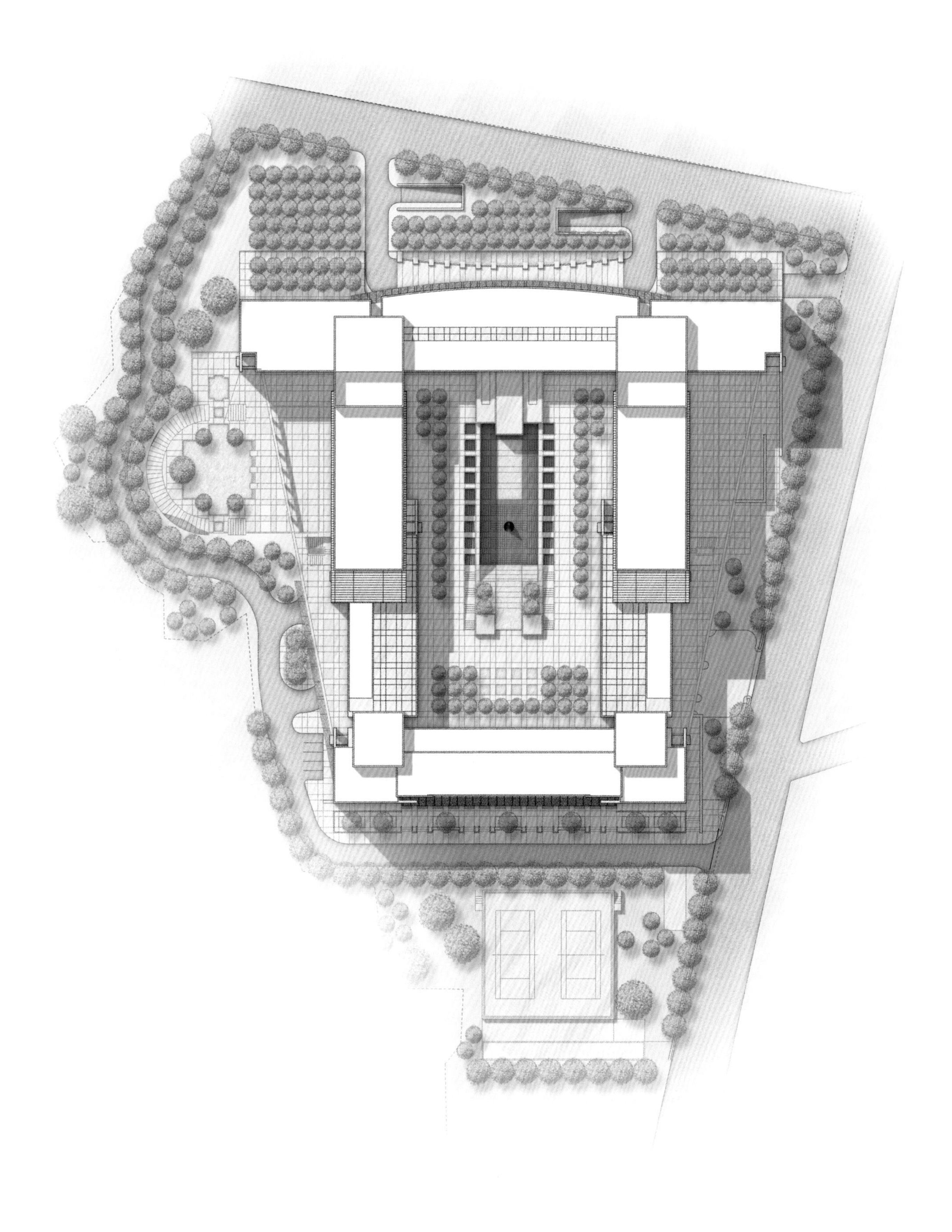

U–505潜水艇展馆
U-505 SUBMARINE EXHIBIT

Chicago, Illinois, USA

This structure is located 40 feet below the northeast lawn of Chicago's Museum of Science and Industry. After being on display outside the museum for 50 years, the German U-505 submarine was severely corroded. Conserving the 700-ton (635-metric-ton) artifact–listed on the National Register of Historic Places–was paramount. The climate-controlled, underground pavilion fully displays the submarine and protects it from further deterioration. The 27,000-square-foot addition includes interactive exhibits, a theater, and a wall commemorating American sailors from World War II. Renovations were made to 5,000 square feet of adjacent space.

The museum's objective was to insert this structure without compromising the historic significance of the original building–one of only a few remaining from the 1893 World's Fair. To meet this goal, the addition was placed below grade, a uniquely appropriate setting for a submarine.

The design recalls World War II–era submarine pens and dry docks, with exposed concrete walls and arched steel girders. Visitors can now walk around the submarine's entire perimeter; a series of ramps provides cantilevered observation points and access on two levels. The structure's unique form responds to the shape of the submarine and the physical forces of the site underground. The angled design resulted in a structure that is approximately 10 feet wider at the roof. Because the top of the exhibit space is broader, major features such as the conning tower, periscopes, and artillery can be viewed easily, and spaciousness is achieved within the enclosure. The arched, structural-steel roof girders support up to 7 feet of soil above and could be erected more quickly than a concrete structure after the submarine was lowered into place.

The most extraordinary installation I have ever seen.

–Juror, Chicago Architecture Foundation
Patron of the Year Awards 2006

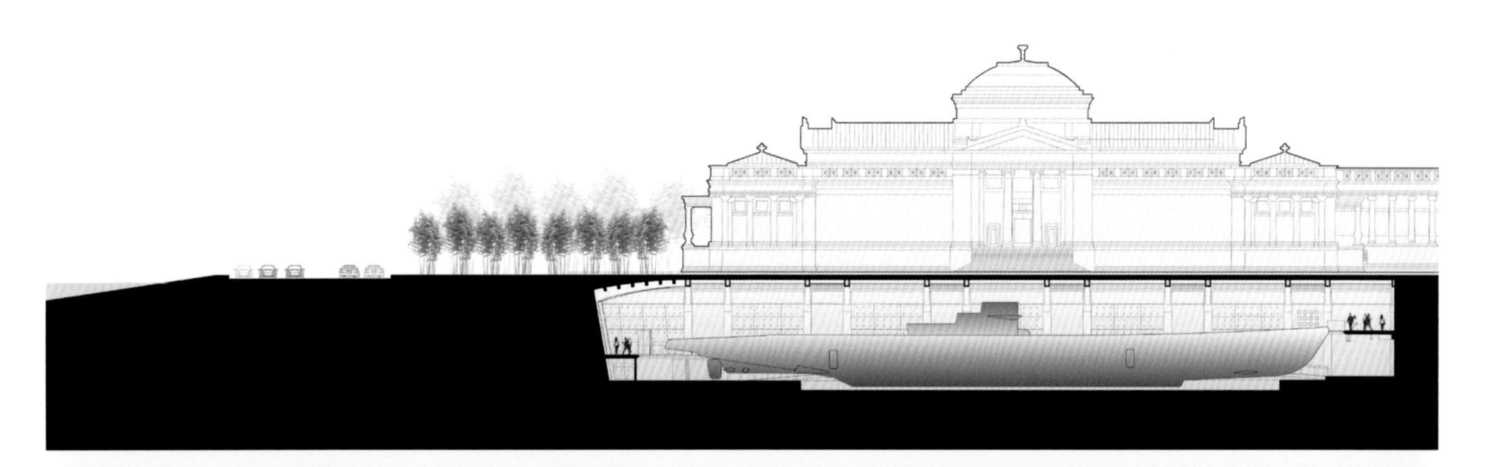

军人球场与伯纳姆公园北侧大楼改建工程
SOLDIER FIELD AND NORTH BURNHAM PARK REDEVELOPMENT

Chicago, Illinois, USA

The original 1924 Soldier Field structure was intended to serve events including track and field, football, auto racing, and public assemblies, but it never fully responded to the different needs of its varied users, becoming home to the Chicago Bears in 1971. Compared to other NFL facilities, the stadium had a poor seating configuration, few amenities, primitive concessions and restrooms, and small, archaic suites. It was an outdated stadium surrounded by 66 acres of paved parking before a joint venture of two firms revitalized the site: Goettsch Partners, with primary responsibility for the master plan and North Burnham Park project, and Wood + Zapata, with primary responsibility for the architectural design of the Soldier Field stadium.

The concept–not replicating the old facility but creating a modern stadium within the confines of the historic structure–juxtaposes the original colonnades against a contemporary steel-and-glass stadium with an asymmetrical shape, featuring four levels of skybox suites and club seating on one side and cantilevered grandstand seating on the other. The long spans and cantilevers used throughout accommodate the existing structure's width and ensure unobstructed views while adding a sense of movement.

The Bears now play on natural turf heated from below, and the team's training and dressing rooms are modernized. Stadium amenities for fans feature added and enhanced restrooms and concessions, 62,000 seats, improved sightlines, two 23- by 82-foot video-replay boards, 133 luxury suites, and three club lounges that overlook the historic colonnades and Lake Michigan. The surrounding 17 acres of landscaped parkland include a veterans sculpture and water wall, a children's garden, a police memorial garden, a winter garden, and a sledding hill. Chicago residents and visitors now enjoy year-round access to the colonnades, a new boulevard for improved vehicular circulation, underground parking, the expanded parkland, and a more cohesive and welcoming campus that connects to the adjacent Field Museum, Shedd Aquarium, and Adler Planetarium.

It's bold. It's unconventional. It looks to the future, while remembering the past. And it works.

–Former Mayor Richard M. Daley, City of Chicago

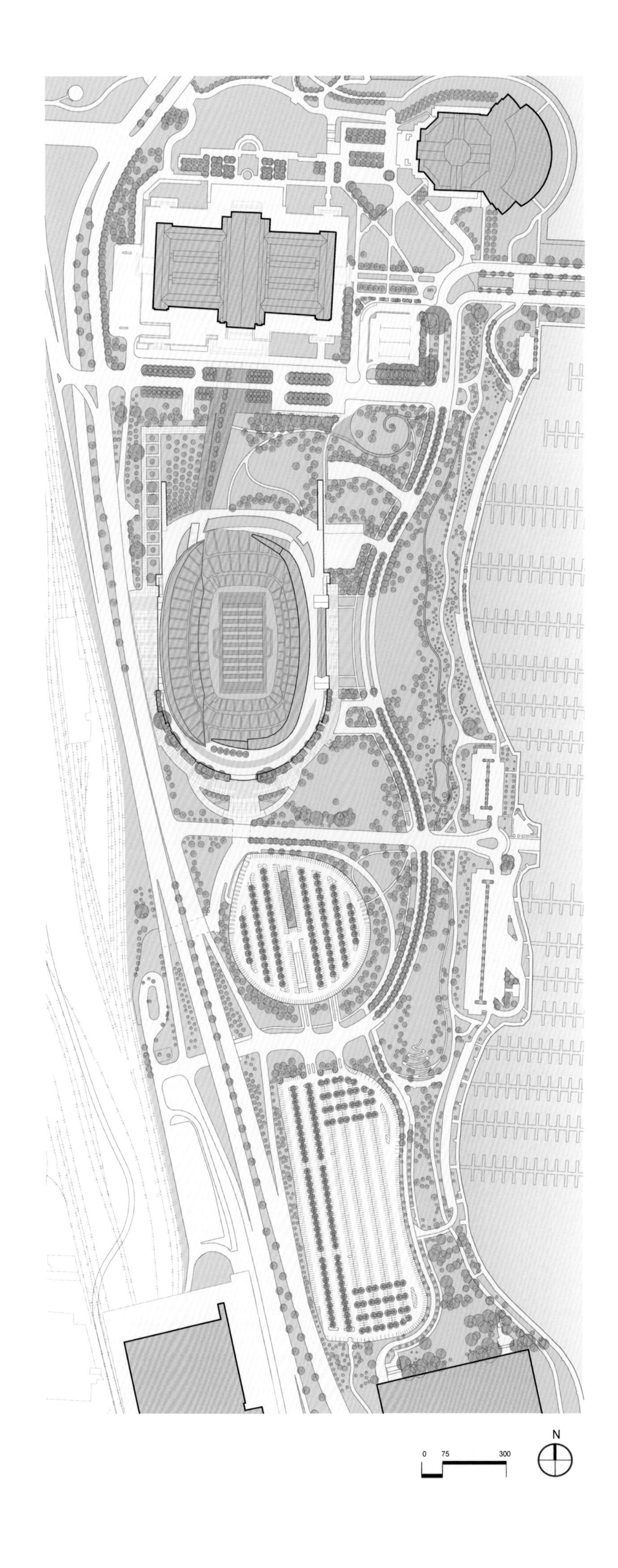
N
0
75
300

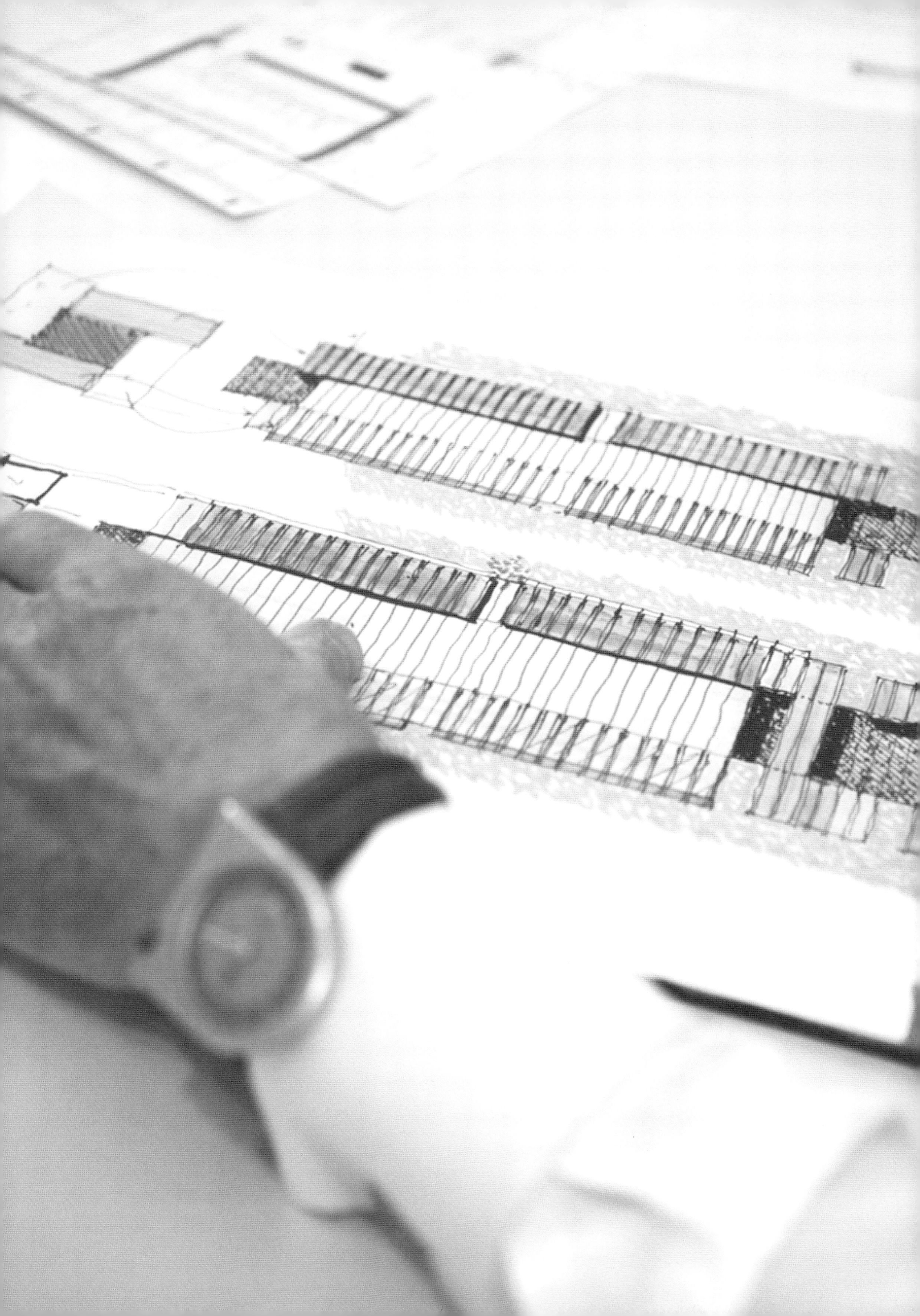

SOUTH
WEST

办公室
OFFICE

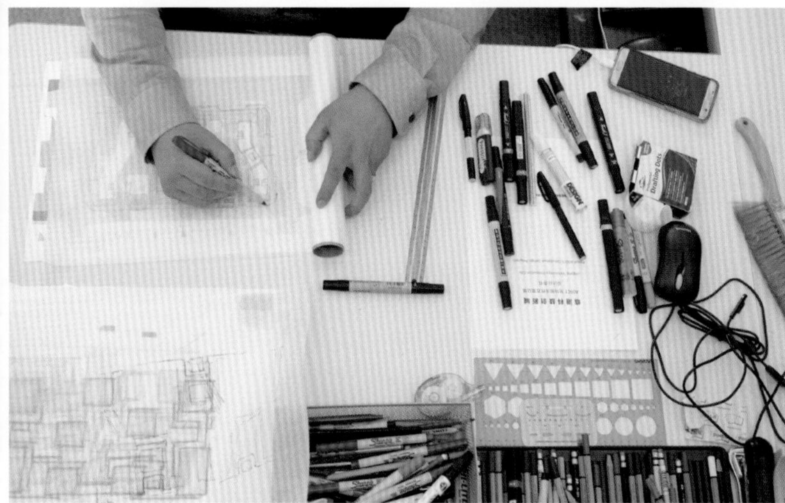

EXIT

项目信息
PROJECT CREDITS

CURRENT

150 NORTH RIVERSIDE
Chicago, Illinois, USA
2011-2017
1,464,000 ft² (136,010 m²)
Client: Riverside Investment & Development
Structural Engineer: Magnusson Klemencic Associates
MEP Consultant: Cosentini Associates
Lighting Designer: One Lux Studio
Landscape Architect: Wolff Landscape Architecture
Photographer: Nick Ulivieri Photography (21); Tom Rossiter Photography (22, 23, 25, 27-31)

ZURICH NORTH AMERICA HEADQUARTERS
Schaumburg, Illinois, USA
2013-2016
783,800 ft² (72,820 m²)
Client: Zurich North America
Developer: CRG Real Estate Solutions
Design-Builder: Clayco
Client Representative, Project Manager & Sustainability: JLL
Structural Engineer: WSP USA
Interior Design & Workplace Strategy: CannonDesign
Lighting Designer: One Lux Studio
Landscape Architect: Hoerr Schaudt Landscape Architects
Civil Engineer: V3 Companies
Sustainability Consultant: Thornton Tomasetti
Photographer: James Steinkamp Photography

AL HILAL BANK OFFICE TOWER
Abu Dhabi, United Arab Emirates
2010-2015
942,600 ft² (87,570 m²)
Client: Al Hilal Bank
Structural Engineer: DeSimone Consulting Engineers
MEP Engineer: Environmental Systems Design
Architect/Engineer of Record: Serex International
Owner's Representative: John Buck International
Lighting Designer: One Lux Studio
Landscape Architect: Wolff Landscape Architecture
Photographer: Lester Ali (45, 46, 50, 51); Tom Rossiter Photography (47-9)

ABU DHABI GLOBAL MARKET SQUARE
Abu Dhabi, United Arab Emirates
2007-2012
5,698,000 ft² (529,360 m²)
Client: Mubadala Real Estate & Infrastructure
Owner's Representative: EC Harris
Structural/MEP Engineer: Oger International
MEP Engineer/Peer Review: Environmental Systems Design
Structural Engineer/Peer Review: Thornton Tomasetti
Architect/Engineer of Record: Serex International
Lighting Designer: One Lux Studio
Landscape Architect: Martha Schwartz Partners
Photographer: Mubadala Real Estate & Infrastructure (53, 59); Lester Ali Photography (54, 55, 57, 60-2); Gerry O'Leary Photography (63)

NANJING QILIN TECHNOLOGY CAMPUS HEADQUARTERS
Nanjing, China
2011-2013
742,700 ft² (69,000 m²)
Client: Guangzhou R&F Properties Co., Ltd.
Associate Architect: Nanjing Architectural Design & Research Institute
Structural/MEP Engineer: Nanjing Architectural Design & Research Institute
Landscape Architect: SWA Group
Photographer: Shen Zhonghai, 1st Image

SOOCHOW SECURITIES HEADQUARTERS
Suzhou, China
2009-2013
441,300 ft² (41,000 m²)
Client: Soochow Securities Co., Ltd.
Associate Architect: Suzhou Industrial Park Design & Research Institute
Structural/MEP Engineer: Suzhou Industrial Park Design & Research Institute
Lighting Designer: One Lux Studio
Landscape Architect: ACLA
Photographer: Shen Zhonghai, 1st Image

R&F YINGKAI SQUARE
Guangzhou, China
2006-2014 (Building), 2016 (Hotel)
1,878,300 ft² (174,500 m²)
Client: Guangzhou R&F Properties Co., Ltd.
Associate Architect: Guangzhou Residential Architectural Design Institute
Structural Engineer: Beijing R&F Properties Development Co.
MEP Engineer: Arup
Lighting Designer: Brandston Partnership Inc. (BPI)
Landscape Architect: ACLA
Hotel Interior Designer: Super Potato
Photographer: Shen Zhonghai, 1st Image

CHICONY SQUARE
Chengdu, China
2005-2011 (Building), 2016 (Hotel)
2,000,000 ft² (185,810 m²)
Client: Chicony Co.
Associate Architect: Architectural Design Institute of Sichuan Province
Structural Engineer: Architectural Design Institute of Sichuan Province
MEP Engineer: Parsons Brinckerhoff / Architectural Design Institute of Sichuan Province
Hotel Interior Designer: Tonychi and Associates
Landscape Architect: P Landscape
Photographer: Shen Zhonghai, 1st Image

GRAND HYATT DALIAN
Dalian, China
2007-2014
1,075,300 ft² (99,900 m²)
Client: China Resources Land Limited (CR Land)
Associate Architect: China Architecture Design & Research Group (CADG)
Structural Engineer: RBS Architectural Engineering Design Associates
MEP Engineer: Meinhardt Group
Interior Designer: LTW Design Works
Rooftop Restaurant Interior Designer: Celia Chu Design
Lighting Designer: Tino Kwan Lighting Consultants
Landscape Architect: ACLA
Photographer: Shen Zhonghai, 1st Image

HOTEL KAPOK SHENZHEN BAY
Shenzhen, China
2009–2012
290,600 ft² (27,000 m²)
Client: China Resources Land Limited (CR Land)
Associate Architect: Beijing Institute of Architectural Design
Structural/MEP Engineer: Beijing Institute of Architectural Design
Interior Designer: Celia Chu Design
Lighting Designer: Beijing Lighting Design
Photographer: Shen Zhonghai, 1st Image

HANGZHOU MARRIOTT HOTEL QIANJIANG
Hangzhou, China
2008–2016
1,005,300 ft² (93,400 m²)
Client: Hangzhou UDC Group
Associate Architect: Architectural & Civil Engineering Design Institute Co.
Structural Engineer: Architectural & Civil Engineering Design Institute Co.
MEP Engineer: Parsons Brinckerhoff
Interior Designer: CCD/Cheng Chune Design
Lighting Designer: Brandston Partnership Inc. (BPI)
Landscape Architect: Place Design Group
Photographer: Shen Zhonghai, 1st Image

VICEROY CHICAGO
Chicago, Illinois, USA
2014–2017
150,000 ft² (13,940 m²)
Client: Convexity Properties
Structural Engineer: Forefront Structural Engineers
MEP Engineer: IMEG
Hotel Interior Designer: TAL Studio
Restaurant Interior Designer: AvroKO
Photographer: James Steinkamp Photography

LONDONHOUSE CHICAGO
Chicago, Illinois, USA
2013–2016
390,000 ft² (36,230 m²)
Client: Oxford Capital Group, LLC; Angelo, Gordon & Co.
MEP Engineer: WMA Consulting Engineers
Interior Designer: Simeone Deary Design Group
Structural Engineer: TGRWA
Exterior Masonry Restoration: Klein & Hoffman
Photographer: Tom Rossiter Photography

THE WRIGLEY BUILDING
Chicago, Illinois, USA
2011–2013
680,000 ft² (63,170 m²)
Client: BDT Capital Partners, LLC; Zeller Realty Group
MEP Engineer: Environmental Systems Design
Structural Engineer: Klein & Hoffman
Plaza Structural Engineer: Wiss, Janney, Elstner Associates
Masonry Façade Engineer: BTC/Building Technology Consultants
Landmark and Tax Consultant: MacRostie Historic Advisors
Photographer: Jon Miller, Hedrich Blessing

PATRICK G. AND SHIRLEY W. RYAN CENTER FOR THE MUSICAL ARTS
Evanston, Illinois, USA
2008–2015
152,000 ft² (14,120 m²)
Client: Northwestern University
Structural Engineer: Thornton Tomasetti
MEP Engineer: Cosentini Associates
Acoustical Consultant: Kirkegaard Associates
Landscape Architect: Hoerr Schaudt Landscape Architects
Lighting Designer: Schuler Shook
Theater Planner: Schuler Shook
Photographer: Tom Rossiter Photography (129–31, 133 top, 135–9); Goettsch Partners (133 bottom)

ON THE BOARDS

110 NORTH WACKER
Chicago, Illinois, USA
2016–2020
1,650,000 ft² (153,290 m²)
Client: The Howard Hughes Corporation / Riverside Investment & Development
Structural Engineer: Thornton Tomasetti
MEP Consultant: Environmental Systems Design
Lighting Designer: One Lux Studio
Landscape Architect: Wolff Landscape Architecture

PARK TOWER AT TRANSBAY
San Francisco, California, USA
2014–2018
803,700 ft² (74,670 m²)
Client: The John Buck Company; Golub & Company LLC; MetLife
Associate Architect: Solomon Cordwell Buenz; Stevens & Associates
Structural Engineer: Magnusson Klemencic Associates
MEP Engineer: WSP
Lighting Designer: Lightswitch Architectural; IA Interior Architects
Landscape Architect: Andrea Cochran Landscape Architecture

DUBAI MIXED-USE TOWER
Dubai, United Arab Emirates
2016
2,439,300 ft² (226,620 m²)
Client: Confidential
Associate Architect: ARC International
Structural/MEP Engineer: WSP

HILTON RIYADH HOTEL & RESIDENCES
Riyadh, Saudi Arabia
2009–2017
1,194,800 ft² (111,000 m²)
Client: General Organization for Social Insurance (GOSI)
Design Partner: Omrania & Associates
Interior Designer: Wrenn Interiors
Structural/MEP Engineer: Omrania & Associates
Lighting Designer: CD+M Lighting Design Group
Landscape Architect: Omrania & Associates

MENNICA LEGACY TOWER
Warsaw, Poland
2014–2019
1,237,800 ft² (115,000 m²)
Client: Golub GetHouse; Mennica Towers GGH MT Sp z o. o. S.K.A.
Associate Architect: Epstein
Structural/MEP Engineer: Epstein
Lighting Designer: One Lux Studio
Landscape Architect: RS Architektura Krajobrazu

项目信息

PROJECT CREDITS *cont.*

SHENZHEN 800 LANDMARK TOWER
Shenzhen, China
2017–2024
6,311,600 ft² (586,370 m²)
Client: Confidential

THE SUMMIT
Suzhou, China
2012–2017
1,615,200 ft² (150,060 m²)
Client: Tishman Speyer
Associate Architect: Suzhou Institute of Architectural Design (SIAD)
Structural/MEP Engineer: WSP
Lighting Designer: KGM International Lighting Design
Landscape Architect: Sasaki Associates
Curtain Wall Consultant: Inhabit Group
Photographer: Shen Zhonghai, 1st Image

AVIVA-COFCO QIANHAI TOWERS
Shenzhen, China
2016–2019
2,045,100 ft² (190,000 m²)
Client: COFCO Group
Associate Architect: Huasen Architectural & Engineering Design Consultants
Structural Engineer: PT International Design Consultants
MEP Engineer: WSP
Curtain Wall Consultant: RFR Consulting Engineers

SHANGHAI LINGANG INTERNATIONAL CONFERENCE CENTER
Shanghai, China
2014–2017 (Phase 1)
3,207,000 ft² (297,960 m²)
Client: Shanghai Lujiazui Xinchen Investment Company
Associate Architect: ECADI
Structural/MEP Engineer: ECADI

QIANHAI DEVELOPMENT
Shenzhen, China
2013–2019
5,076,300 ft² (471,600 m²)
Client: China Resources Land Limited (CR Land)
Associate Architect: Architectural and Research Institute of Guangdong Province (GDI)
Retail Architect: Benoy
Structural Engineer: WSP
MEP Engineer: Meinhardt Group
Hotel Interior Designer: Hanya Design
Lighting Designer: Brandston Partnership Inc. (BPI)
Landscape Architect: ACLA
Curtain Wall Consultant: WSP

CHANCHENG GREENLAND CENTER
Foshan, China
2012–2016 (Phase 1), 2018 (Phase 2)
4,693,100 ft² (436,000 m²)
Client: Greenland Group Guangdong Real Estate Department
Associate Architect: Architectural Design and Research Institute of Guangdong Province
Structural Engineer: Architectural Design and Research Institute of Guangdong Province
MEP Engineer: BON Engineering Consultants
Curtain Wall Consultant: SuP Ingenieure GmbH

CHANCHENG GREENLAND CENTER–PHASE 2 COMPETITION
Foshan, China
2014
6,127,900 ft² (569,300 m²)
Client: Greenland Group Guangdong Real Estate Department
Associate Architect: Architectural Design and Research Institute of Guangdong Province
Structural Engineer: Architectural Design and Research Institute of Guangdong Province
Structural Consultant: WSP USA
MEP Engineer: BON Engineering Consultants

CHINA HOROY QIANHAI GUANZE MIXED-USE
Shenzhen, China
2014–2018
3,444,500 ft² (320,000 m²)
Client: Horoy Qianhai International Holdings Limited
Associate Architect: Architectural and Research Institute of Guangdong Province (GDI)
Retail Architect: Lead 8
Structural Engineer: Arup
MEP Engineer: WSP
Hotel Interior Designer: Yabu Pushelberg
Hotel Restaurants/Spa and Fitness Interior Designer: Design Studio Spin
Lighting Designer: DASUN
Landscape Architect: SWA Group
Curtain Wall Consultant: Inhabit Group

GREENLAND OPTICS VALLEY CENTER
Wuhan, China
2016–2022
3,390,600 ft² (315,000 m²)
Client: Greenland Group
Associate Architect: China Architectural Design & Research Group (CADG)
Structural Engineer: Thornton Tomasetti
MEP Engineer: WSP

NANNING CHINA RESOURCES CENTER TOWER
Nanning, China
Design/Completion: 2011–2019
2,930,600 ft² (272,260 m²)
Client: China Resources Land Limited (CR Land)
Associate Architect: CCDI Group
Structural Engineer: RBS Architectural Engineering Design Associates
MEP Engineer: WSP
Hotel Interior Designer: Studio Munge
Lighting Designer: Han Design Associates (HDA)
Landscape Architect: ADI
Curtain Wall Consultant: WSP

ROSEWOOD SANYA AND SANYA FORUM
Sanya, China
2012–2017
1,630,700 ft² (151,500 m²)
Client: Poly Real Estate (Group) Co., Ltd.
Associate Architect: Guangzhou Design Institute
Structural Engineer: RBS Architectural Engineering Design Associates
MEP Engineer: Parsons Brinckerhoff
Hotel Interior Designer: AB Concept
Apartment Interior Designer: Spaces
Forum Interior Designer: Citygroup
Lighting Designer: Project Lighting Design (PLD)
Landscape Architect: P Landscape
Curtain Wall Consultant: Suma Façade Consultants (SFC)
Photographer: Chen Ji

LEGACY

111 SOUTH WACKER
Chicago, Illinois, USA
2000–2005
1,457,000 ft² (135,360 m²)
Client: The John Buck Company
Structural Engineer: Magnusson Klemencic Associates
Mechanical Engineer: WMA Consulting Engineers
MEP Consultant: Environmental Systems Design
Lighting Designer: One Lux Studio
Photographer: James Steinkamp Photography

300 EAST RANDOLPH
Chicago, Illinois, USA
1994–1997 (Phase 1), 2010 (Phase 2)
2,350,000 ft² (218,320 m²)
Client: Health Care Service Corporation
Developer (Phase 1): Walsh Higgins & Company
Development Manager (Phase 2): The John Buck Company
Structural Engineer (Phase 1): Christopher P. Stefanos Associates
Structural Engineer (Phase 2): Magnusson Klemencic Associates
MEP Engineer: Cosentini Associates
Lighting Designer: One Lux Studio
Photographer: James Steinkamp Photography (219, 220 right, 221, 225); Goettsch Partners (220 left); Marshall Gerometta (220 middle); Hedrich Blessing (222–3, 224)

UBS TOWER
Chicago, Illinois, USA
1998–2001
1,754,000 ft² (162,950 m²)
Client: The John Buck Company
Structural Engineer: Thornton Tomasetti
MEP Engineer: Environmental Systems Design
Lighting Designer: One Lux Studio
Landscape Architect: Peter Walker & Partners
Photographer: Jon Miller, Hedrich Blessing (227–30); David B. Seide, Defined Space (231)

CHARLES SQUARE CENTER
Prague, Czech Republic
1997–2002
290,500 ft² (26,990 m²)
Client: GE Capital Golub Europe, LLC
Associate Architect: SIAL s.r.o.
Structural Engineer: Thornton Tomasetti
MEP Engineer: Building Services Group
Lighting Designer: One Lux Studio
Photographer: Pavel Štecha

CHINA DIAMOND EXCHANGE CENTER
Shanghai, China
2005–2009
535,500 ft² (49,750 m²)
Client: Shanghai Lujiazui Development Co., Ltd.
Associate Architect: Shanghai Zhong-fu Architects
Structural Engineer: Shanghai Tong-qing Technologic Development
MEP Engineer: Shanghai Zhong-fu Architects
Lighting Designer: Shanghai New Century Co.
Landscape Architect: ADI Limited
Photographer: Shen Zhonghai, 1st Image

GRAND HYATT SÃO PAULO
São Paulo, Brazil
1998–2002
592,000 ft² (55,000 m²)
Client: Sociedad Latinoamericana de Inversiones S.A.
Associate Architect: Escritório Técnico Júlio Neves S/C
Structural Engineer: AHFsa Civil Engineers
MEP Engineer: Cosentini Associates
Interior Designer: Remedios Siembieda
Landscape Architect: SWA Group/Isabel Duprat
Photographer: Eduardo Girão, Estúdiogirão

GRAND HYATT MUMBAI
Mumbai, India
1996–2004
1,200,000 ft² (111,480 m²)
Client: Unison Hotels Ltd.
Associate Architect: Gherzi Eastern Limited
Structural Engineer: Gherzi Eastern Limited
MEP Engineer: Spectral Services Consultants
Interior Designer: Remedios Siembieda
Landscape Architect: SWA Group
Photographer: Tom Fox, SWA Group (245, 247 top); Goettsch Partners (247 bottom)

U-505 SUBMARINE EXHIBIT
Chicago, Illinois, USA
2001–2005
32,000 ft² (2,970 m²)
Client: Museum of Science and Industry
Program Manager: Jones Lang LaSalle
Structural Engineer: Halvorson + Partners
MEP Engineer: Primera Engineers
Lighting Designer: Available Light
Photographer: Mark Ballogg, Steinkamp/Ballogg Photography (249, 250); David B. Seide, Defined Space (251)

SOLDIER FIELD AND NORTH BURNHAM PARK REDEVELOPMENT
Chicago, Illinois, USA
1999–2003
Additional Parkland: 17 acres (6.88 hectares)
Stadium: 1,600,000 ft² (148,640 m²)
Client: Chicago Bears
Owner: Chicago Park District
Landscape Architect: Peter Lindsay Schaudt Landscape Architecture
Structural Engineer: Thornton Tomasetti
MEP Engineer: Ellerbe Becket
Mechanical/Electrical Engineer: Environmental Systems Design
Civil Engineer: V3 Consultants
Graphics and Wayfinding: Catt Lyon Design
Lighting Designer: Schuler Shook
Photographer: David B. Seide, Defined Space (253, 255 bottom, 257); Doug Fogelson (255 top, 256)

Additional Photography: Michelle Litvin (except as noted), Barb Levant Photo (7 top left), Tom Rossiter Photography (18–19), Omrania & Associates (140–1), Mark Ballogg, Steinkamp/Ballogg Photography (210–11), Steve Ewert Photography (262 top left, 263 middle left, 264 top right, 265 middle right and bottom right)

致谢
ACKNOWLEDGEMENTS

The partners would like to first thank our clients, without whom the work depicted within this monograph would not have been possible. They have provided us with rich and varied challenges, and have been our collaborators throughout. Each client has brought to us a unique opportunity, and we have endeavored to exceed their expectations. It has been our honor and privilege to work with them.

Our projects require the involvement of the highest quality consultants and contractors, and we gratefully acknowledge these key members of the building team.

We would also like to recognize all current and past staff for their contributions. The success of these projects is a direct result of a shared spirit and appreciation of architecture.

Finally, Goettsch Partners would like to thank our dedicated, enthusiastic and talented staff, all of whom at one point or another have made contributions to this comprehensive presentation of the work of our firm. We are particularly grateful to the efforts of Paul De Santis, Matt Larson, William Netter and Stephanie Pelzer, who all committed a significant amount of time and energy over several months to ensure the quality and accuracy of the content presented.

索引
INDEX